3rd International Conference on

MICROWAVE, OPTICAL AND COMMUNICATION ENGINEERING

(ICMOCE-2025)

3rd International Conference on

MICROWAVE, OPTICAL AND COMMUNICATION ENGINEERING

(ICMOCE-2025)

Book of Abstracts

Editors

DEBALINA GHOSH

PRASANT KUMAR SAHU

INDIAN INSTITUTE OF TECHNOLOGY (IIT) BHUBANESWAR

Argl–Jatni Road, Kansapada, Odisha 752050

B L O O M S B U R Y

NEW DELHI • LONDON • OXFORD • NEW YORK • SYDNEY

BLOOMSBURY INDIA
Bloomsbury Publishing India Pvt. Ltd
Second Floor, LSC Building No. 4, DDA Complex,
Pocket C - 6 & 7, Vasant Kunj,
New Delhi 110070

BLOOMSBURY, BLOOMSBURY QUEST and the Diana logo are trademarks of Bloomsbury Publishing Plc

First published in India 2025
This edition published 2025

ISBN: 978-93-69528-25-7
2 4 6 8 10 9 7 5 3 1

Printed and bound in India by Replika Press Pvt. Ltd

CONTENTS

MESSAGES

Message from Director, IIT Bhubaneswar

I am delighted to learn that the School of Electrical and Computer Sciences at the Indian Institute of Technology Bhubaneswar is organizing the 3rd edition of the International Conference on Microwave, Optical and Communication Engineering (ICMOCE-2025) during 23rd to 25th May 2023. The ICMOCE-2025 will provide a global forum for researchers, students, developers, faculty, scientists, and engineers from industry, academia, and research and development organizations to present and share their pioneering technical work and leading-edge research. The conference will inspire young researchers and provide professionals and scientists with new knowledge.

I congratulate all involved in organizing this conference and wish the conference a grand success.

Best regards,

Shreepad Karmalkar
Director, IIT Bhubaneswar

Message from Dean Continuing Education, IIT Bhubaneswar

It is my great pleasure to extend a warm welcome to all participants of this esteemed International Conference on Microwave, Optical and Communication Engineering 2025. As the Dean of Continuing Education, I am honored to be part of this remarkable gathering of scholars, researchers, industry experts, and students from India and around the world.

This conference is a vital platform for the exchange of cutting-edge knowledge, the forging of new partnerships, and the nurturing of ideas that will shape the future of our fields. I extend my heartfelt congratulation to the organizing committee for their tireless efforts in organising this conference. I thank all contributors for their valuable insights and dedication. I hope that your experience here is intellectually enriching, professionally rewarding, and personally inspiring.

Wishing you a successful and engaging conference.

Prof. V. Pandu Ranga
Dean Continuing Education

Message from Conference General Chair

It is my great pleasure to welcome you all to the 3rd *International Conference on Microwave, Optical and Communication Engineering*. As the Conference General Chair, I am honoured to have had the opportunity to work with a dedicated team of organizers and reviewers to bring this event to fruition.

This conference has brought together experts from diverse fields including microwave, optical, communication, signal processing, data science, and bio sensing to share their latest research and ideas. The selected papers represent high quality, innovative contributions from leading researchers, displaying the breadth and depth of work in these areas.

I extend my sincere thanks to the authors, reviewers, and organizing team members for their dedication and hard work. The success of this conference would not have been possible without the hard work and dedication of our authors, reviewers, and organizers. I extend my sincerest gratitude to everyone involved in making this conference a success.

I hope that the papers presented in this conference will inspire future research, foster new collaborations, and contribute to the advancement of knowledge in the themed area and allied areas.

Thank you for your participation and contribution to this conference, and I look forward to seeing you again at future events.

Warm regards,

Dr. Debalina Ghosh
Associate Professor,
Microwave and Optical Engineering Research Group (MOE),
School of Electrical and Computer Sciences,
IIT Bhubaneswar

Message from Conference Organising Secretary

I welcome you to this International Conference on Microwave, Optical and Communication Engineering (ICMOCE-2025) and thank you for being part of this important academic and professional gathering. This conference brings together a vibrant community of researchers, academicians, industry professionals, and students from across the Institutions in India and abroad, exchange ideas, and explore innovative solutions to the challenges of our time. Organizing this event has been a journey of learning and teamwork. I extend my heartfelt appreciation to the members of the organizing committee, advisory committee, the technical reviewers, our sponsors, and all those who contributed their time and expertise to make this conference possible.

We hope the sessions prove to be thought provoking. May this conference serve not only as a platform for academic advancement but also as a medium for future collaborations and innovations. We look forward to welcoming you at our future events.

Warm regards,

Dr. Prasant Kumar Sahu
Professor
School of Electrical and Computer Sciences
IIT Bhubaneswar

ICMOCE-2025 COMMITTEES

Chief Patron

Director, IIT Bhubaneswar

Patron

Dean CE, IIT Bhubaneswar

Convener

Prof. D. Ghosh, IIT Bhubaneswar

Organizing Secretary

Prof. P. K. Sahu, IIT Bhubaneswar

Advisory Committee

- Prof. Brian Cullum, University of Maryland Baltimore County, USA
- Prof. Yeon Ho Chung, Pukyong National University, Korea (South)
- Prof. Christophe, Peucheret, University of Renes, France
- Prof. Xiaoying Zhuang, Leibniz University Hannover, Germany
- Prof. Abhirup Das Barman, Dept. of Radio Physics & Electronics, CU, India
- Prof. Dharmendra Singh, IIT Roorkee, India
- Prof. Ganapati Panda,C V Raman Global University-Odisha
- Prof. Animesh Biswas, IIT Kanpur, India
- Prof. Partha Roop, University of Auckland, New Zealand
- Prof. Parthasarathi Chakrabarti, IIT BHU, India
- Prof. A R Harish, IIT Kanpur, India
- Prof. Vikram Gadre, IIT Bombay, India
- Prof. Krishnamachar Prasad, Auckland University of Technology, NZ
- Prof. Debashis Gosh, IIT Roorkee, India
- Prof. Rohit Sinha, IIT Guwahati, India
- Prof. Arun Somani, College of Engineering, Iowa State University, USA
- Dr. Ashutosh Kedar, LRDE, DRDO, India

ICMOCE-2025 Committee Chairs

- **General Chair**
  Dr. Debalina Ghosh
- **Technical Program Chairs**
  Dr. Prasant Kumar Sahu & Dr. S K Varshney

- **Microwave Track Chairs**
  Dr. Debapratim Ghosh & Dr. Debalina Ghosh
- **Optical Track Chairs**
  Dr. Rajan Jha & Dr. Prasant Kumar Sahu
- **Communication Track Chairs**
  Dr. Siddartha Borkotoky & Dr. Atri Mukhopadhyay
- **Biosensing Track Chairs**
  Dr. Niladri Bihari Puhan & Dr. Himanshu P Padole
- **Data Science and AI Track Chairs**
  Dr. Manoranjan Satpathy, Dr. Abhik Jana & Dr. Ram P. Padhy
- **Publicity Chairs**
  Dr. Pravas R Sahu & Dr. Vijay Shankar Rao
- **IEEE Region 10 Outreach Chair**
  Dr. Subhranshu R Samantaray
- **Local Arrangements Chairs**
  Dr. Barathram Ramkumar, Dr. Soumya P. Das & Dr. Navjeet Bagga
- **Publications Chair**
  Dr. Debalina Ghosh
- **Finance Chair**
  Dr. Prasant Kumar Sahu
- **Social Programs and Hospitality Chairs**
  Dr. Akshay K. & Dr. Sayan Dey
- **Secretariat**
  Mr. Mrinal Datta

ICMOCE-2025 Technical Program Committee (TPC)

Prof. A Agrawal	VNIT Nagpur
Prof. A K Singh	IIT (BHU) Varanasi
Prof. Debashis Ghosh	IIT Roorkee
Prof. A V Nandedkar	SGGS Nanded
Prof. Amarjit Kumar	NIT Warangal
Prof. Anoop Thomas	IIT Bhubaneswar
Prof. Arun Kumar Gande	NIT Warangal
Prof. Arvind Kumar	VNIT Nagpur
Prof. B Ramkumar	IIT Bhubaneswar
Prof. B Rebekka	NIT Trichy
Prof. Bibhudendra Acharya	NIT Raipur

Prof. Bijayananda Patnaik	NIT Raipur
Prof. D Sriram Kumar	NIT Trichy
Prof. D P Dogra	IIT Bhubaneswar
Prof. Debalina Ghosh	IIT Bhubaneswar
Prof. Debapratim Ghosh	IIT Bhubaneswar
Prof. Dileep K Upadhyay	BIT Mesra
Prof. Gautam K Mahanti	NIT Durgapur
Prof. Harish Sahoo	VSSUT Burla
Prof. Hemant Kumar	NIT Trichy
Prof. Himansu S Pradhan	NIT Warangal
Prof. Jagana Bihari Padhy	GIET Gunupur
Prof. Lalit Kumar	NIT Calicut
Prof. Laxman Pappula	K L University
Prof. M R Saptahy	IIT Bhubaneswar
Prof. M Ravi Kumar	MNIT Jaipur
Prof. M S Manikandan	IIT Palakkad
Prof. Mahesh Sonth	CMRTC Hyderabad
Prof. Maheswaran P	NIT Trichy
Prof. Mandal Durbadal	NIT Durgapur
Prof. Abhik Jana	IIT Bhubaneswar
Prof. Nitya Tiwari	IIT Bhubaneswar
Prof. Himanshu Padole	IIT Bhubaneswar
Prof. Navjeet Bagga	IIT Bhubaneswar
Prof. Ayan Palchadhuri	IIT Bhubaneswar
Prof. Shilpi R Kerketta	IISc Bangalore
Prof. Sayan De	IIT Bhubaneswar
Prof. Santoshkumar H	JSPM University Pune
Prof. R S Deshpande	JSPM University Pune
Prof. Aditya Mehra	JSPM University Pune
Prof. Manish Okade	NIT Rourkela
Prof. Sidhrath Desmukh	NIT Raipur
Prof. Saikat Majumder	NIT Raipur
Prof. Anshul Gupta	NIT Raipur
Prof. Mihir Kumar Mohanty	SOA University
Prof. N B Puhan	IIT Bhubaneswar
Prof. N M Biradar	BKIT Bhalki
Prof. N V L N Murthy	IIT Tirupati

Prof. Nithin V George	IIT Gandhinagar
Prof. P K Sahu	IIT Bhubaneswar
Prof. P R Sahu	IIT Bhubaneswar
Prof. P Rath	IIT Bhubaneswar
Prof. Prakash Kumar Panda	SOA University
Prof. Pyari Mohan Pradhan	IIT Roorkee
Prof. Rajan Jha	IIT Bhubaneswar
Prof. Rajat Kumar Giri	IIIT Bhubaneswar
Prof. Ratnakar Dash	NIT Rourkela
Prof. Ravi Kumar Gangwar	ISM Dhanbad
Prof. S Raghavan	NIT Trichy
Prof. S Anuradha	NIT Warangal
Prof. S K Behera	NIT Rourkela
Prof. S K Varshney	IIT Kharagpur
Prof. S P Das	IIT Bhubaneswar
Prof. S S Jagannath Nanda	MNIT Jaipur
Prof. Sanjay Gowre	BKIT Karnataka
Prof. Santanu Kumar Behera	NIT Rourkela
Prof. Siddhartha Borkotoky	IIT Bhubaneswar
Prof. Sitanshu Sekhar Sahu	BIT Mesra
Prof. Subhrajit Pradhan	GIET Gunupur
Prof. Subrata Maiti	NIT Rourkela
Prof. Sudipta Mahapatra	IIT Kharagpur
Prof. Surya Prakash Singh	IIIT Bhubaneswar
Prof. Urmila Bhanja	IGIT Sarang
Prof. Viswanath K	SIT Tumkur
Prof. Padmalochan Bera	IIT Bhubaneswar
Prof. Nijwm Wary	IIT Bhubaneswar
Prof. Srinivas Boppu	IIT Bhubaneswar
Prof. Veerendra Dakulagi	GNDEC Bidar
Prof. Hemanta Sahu	VIT Vellore
Prof. Mrutyunjay Rout	NIT Jamsedpur
Prof. B.Malarkodi	NIT Trichy
Prof. Suryakanth Bardabadi	BKIT, Bhalki
Prof. Shubhangi D C	VTU, Kaliburagi
Prof. D P Achraya	NIT Rourkela
Prof. M Bhaskar	NIT Trichy

Prof. Laxmi Bhar Biswal	JSPN University Pune
Prof. Rohit Sinha	IIT Guwahati
Prof. Navjeet Bagga	IIT Bhubaneswar
Prof. Atri Mukhopadhyay	IIT Bhubaneswar
Prof. Himanshu P Padole	IIT Bhubaneswar
Prof. Arun Somani	Iowa State University

ICMOCE-2025 Reviewers' Committee

Prof. Abhik Jana	IIT Bhubaneswar
Prof. Abhisek Dash	Max Planck Institute
Prof. Ajay Dash	IIIT Bhubaneswar
Prof. Akshay Ojha	IIT Bhubaneswar
Prof. Alapan Kuila	IIT KHARAGPUR
Prof. Amit Ranjan Azad	IIIT Kalyani
Prof. Anjali Mohapatra	IIIT Bhubaneswar
Prof. Ankan Mullick	IIT Kharagpur
Prof. Anoop Thomas	IIT Bhubaneswar
Prof. Ashwini Sawant	IIT Guwahati
Prof. Atri Mukhopadhyay	IIT Bhubaneswar
Prof. Ayan Palchaudhuri	IIT Bhubaneswar
Prof. Barathram Ramkumar	IIT Bhubaneswar
Prof. Bharati Mishra	IIIT Bhubaneswar
Prof. Bibhudatta Sahoo	NIT Rourkela
Prof. Bibhudendra Acharya	NIT Raipur
Prof. Bijayananda Patnaik	NIT Raipur
Prof. Chittaranjan Nayak	SRMIST, Kattankulathur
Prof. Debalina Ghosh	IIT Bhubaneswar
Prof. Debapratim Ghosh	IIT Bhubaneswar
Prof. Debi Dogra	IIT Bhubaneswar
Prof. Debiprasad Acharya	NIT Rourkela
Prof. Deepika Agrawal	NIT Raipur
Prof. Dharmendra Dixit	MNNIT Allahabad
Prof. Dileep Upadhyay	BIT, Mesra, Ranchi
Prof. Diptiranjan Samantaray	University of Delhi
Prof. G. Shrikanth Reddy	IIT Mandi
Prof. Gautam K Mahanti	NIT Durgapur

Prof. Harish K. Sahoo	VSSUT Burla
Prof. Hemanta K. Sahu	VIT
Prof. Himanshu Padole	IIT Bhubaneswar
Prof. Himansu Pradhan	NIT Warangal
Prof. Jitendra Rout	NIT Raipur
Prof. Jogesh Dash	NIT Rourkela
Prof. Kamalakar Thakare	KIST, South Korea
Prof. Kuntal Deka	IIT Guwahati
Prof. Lakshman Pappula	GITAM University
Prof. Lalit Kumar	NIT Calicut
Prof. Sabarimalai Manikandan	IIT Palakkad
Prof. Madhukrishna Priyadarsini	NIT Raipur
Prof. Manoranjan Satpathy	IIT Bhubaneswar
Prof. Niladri Puhan	IIT Bhubaneswar
Prof. Nilanjan Biswas	NIT Durgapur
Prof. Nithin George	IIT Gandhinagar
Prof. Nitin Bhatia	IIT Jodhpur
Prof. Nitya Tiwari	IIT Bhubaneswar
Prof. Omveer Sharma	IIT Bhubaneswar
Prof. Padmalochan Bera	IIT Bhubaneswar
Prof. Pradeep Das	NIT Warangal
Prof. Pragati Shrivastava	IIT Bhubaneswar
Prof. Prakash Panda	BPUT Rourkela
Prof. Prasant K. Sahu	IIT Bhubaneswar
Prof. Pravas Sahu	IIT Bhubaneswar
Prof. Praween Nishad	IIT Roorkee
Prof. Punyatoya Routray	IIT Bhubaneswar
Prof. Pyari Pradhan	IIT Roorkee
Prof. Rajan Jha	IIT Bhubaneswar
Prof. Rajat Giri	IIIT Bhubaneswar
Prof. Rakesh Ch. Balabantaray	IIIT Bhubaneswar
Prof. Raktim Haldar	IIT Bhubaneswar
Prof. Ram Padhy	IIT Bhubaneswar
Prof. Ratnakar Dash	NIT Rourkela

Prof. Ravi Kumar Maddila	MNIT Jaipur
Prof. Rinkee Chopra	IIT Indore
Prof. Rohit Verma	Intel Labs
Prof. Rupaban Subadar	NEHU, Shillong
Prof. Saba Fatima	JSPM University, Pune
Prof. Sabyasachi Patra	IIIT Bhubaneswar
Prof. Sai Vinod Vangavolu	CVS Pharmacy
Prof. Saikat Majumder	NIT-Raipur
Prof. Sandipan Sikdar	Leibniz University
Prof. Sanjaykumar Gowre	JSPM, Pune
Prof. Sanjeev Raghuwanshi	IIT (ISM) Dhanbad
Prof. Sanjeev Sharma	IIT Varanasi
Prof. Santanu Behera	NIT Rourkela
Prof. Santosh Behera	NIT Calicut
Prof. Sarang Pendharker	IIT Kharagpur
Prof. Satyasai J. Nanda	MNIT, Jaipur
Prof. Shilpi Kerketta	IIT Bhubaneswar
Prof. Shivani Tripathy	IIT Bhubaneswar
Prof. Shreetam Behera	IIT Bhubaneswar
Prof. Siddharth Deshmukh	NIT Raipur
Prof. Siddhartha Borkotoky	IIT Bhubaneswar
Prof. Sitanshu S. Sahu	BITS Mesra,
Prof. Soumajit Pramanik	IIT Bhilai
Prof. Soumi Das	MPI-SWS
Prof. Sujit Das	NIT Warangal
Prof. Surya Singh	IIIT Bhubaneswar
Prof. Sushree Behera	IIIT Bangalore
Prof. Tirath Sahu	NIT Raipur
Prof. Uday Khankhoje	IIT Madras
Prof. Urmila Bhanja	IGIT
Prof. Veerendra Dakulagi	GNDEC, Bidar
Prof. Vikram M Gadre	IIT Bombay
Prof. Visvanathan Kapinaiah	SIT Tumkur
Prof. Vivek Venkataraman	IIT Delhi

ICMOCE-2025 Volunteer Committee

Student Volunteer Core Committee

- **Ms. Nishtha Shrivastava**, SECS, IIT Bhubaneswar
- **Mr. T. Shubham Rao**, SECS, IIT Bhubaneswar
- **Mr. Vaibhav**, SECS, IIT Bhubaneswar
- **Mr. Sanjeet Kumar Bhagat**, SECS, IIT Bhubaneswar
- **Ms. Amisha Verma**, SECS, IIT Bhubaneswar
- **Mr. Shreepreet Sahu**, SECS, IIT Bhubaneswar
- **Mr. Anoop Pandey**, SECS, IIT Bhubaneswar
- **Ms. Tasmin Mishra**, SECS, IIT Bhubaneswar
- **Mr. Subrat Sahu**, SBS, IIT Bhubaneswar
- **Mr. Satyajit Murmu**, SBS, IIT Bhubaneswar
- **Mr. Pratik Mishra**, SBS, IIT Bhubaneswar
- **Mr. Arvind Kumar Maurya**, SBS, IIT Bhubaneswar
- **Mr. Sushil Kumar**, SECS, IIT Bhubaneswar
- **Ms. Subhrajyoti Sunani**, SECS, IIT Bhubaneswar
- **Mr. Kahrrthikeyaa V**, SES, IIT Bhubaneswar
- **Mr. Nikhil Niranjan**, SECS, IIT Bhubaneswar
- **Mr. Munasa Yuvaraju**, SECS, IIT Bhubaneswar
- **Mr. Priyanshu Mathur**, SECS, IIT Bhubaneswar

Technical Volunteer Core Committee

- Mr. Mrinal Datta, SECS, IIT Bhubaneswar
- Mr. Birata K Nanda, SECS, IIT Bhubaneswar
- Mr. Sk. Tajuddin Ahmed, SECS, IIT Bhubaneswar
- Mr. Ratik Mishra, SECS, IIT Bhubaneswar
- **Ms. Madhusmita D Mohapatra**, SECS, IIT Bhubaneswar
- Mr. Jami Naveen Kumar, SECS, IIT Bhubaneswar
- Mr. Tileswar Mahto, SECS, IIT Bhubaneswar
- Mr. Rabinson Behera, CITSC, IIT Bhubaneswar
- Mr. Ranjit Rao, CITSC, IIT Bhubaneswar
- Mr. Alok Singh, CITSC, IIT Bhubaneswar
- Mr. Praveen Kumar R., CITSC, IIT Bhubaneswar

SESSION CHAIRS

- **Prof. A. K. Ojha**, IIT Bhubaneswar
- **Prof. A. R. Harish**, IIT Kanpur
- **Dr. Abhik Jana**, IIT Bhubaneswar
- **Dr. Ajit Jha**, UiA, Norway
- **Dr. Atri Mukhopadhyay**, IIT Bhubaneswar
- **Prof. B. Sundarajan**, IISc Bangalore
- **Dr. Bijayananda Patnaik**, NIT Raipur
- **Prof. D. P. Acharya**, NIT Rourkela
- **Dr. Debalina Ghosh**, IIT Bhubaneswar
- **Dr. D. P. Dogra**, IIT Bhubaneswar
- **Dr. E. Sivasankar**, NIT Trichy
- **Dr. Himanshu Padole**, IIT Bhubaneswar
- **Dr. Satyasai J.Nanda** , NIT Jaipur
- **Prof. Mihir N Mohanty**, SoA, Bhubaneswar
- **Dr. N. B. Puhan**, IIT Bhubaneswar
- **Dr. Nitya Tiwari**, IIT Bhubaneswar
- **Prof. P. Chakrabarty**, IIT Bhubaneswar
- **Prof. P. R. Sahu**, IIT Bhubaneswar
- **Dr. R. P. Padhy**, IIT Bhubaneswar
- **Prof. Rajan Jha**, IIT Bhubaneswar
- **Dr. Rajesh Kumar**, IIT Roorkee
- **Dr. Rakesh Balabantaray**, IIIT Bhubaneswar
- **Dr. Ravi K. Maddila**, NIT Jaipur
- **Dr. S. P. Dash**, IIT Bhubaneswar
- **Dr. Sanjay K. Gowre**, JSPM, Pune
- **Dr. Siddhartha Borkotoky**, IIT Bhubaneswar
- **Dr. Sourabh Pandey**, IIT Patna
- **Dr. Srinivasarao C.**, NIT Warangal
- **Dr. Subhash Chander**, SSPL, DRDO
- **Dr. Varun Kumar**, Shimane University, Japan
- **Dr. Veerendra D.**, GNDEC, Bider
- **Prof. S. Mahapatra**, IIT Kharagpur
- **Prof. Shiva Kumar**, McMaster University, Canada
- **Prof. Vikram Gadre**, IIT Bombay

TECHNICAL PROGRAM SCHEDULE FOR ICMOCE-2025

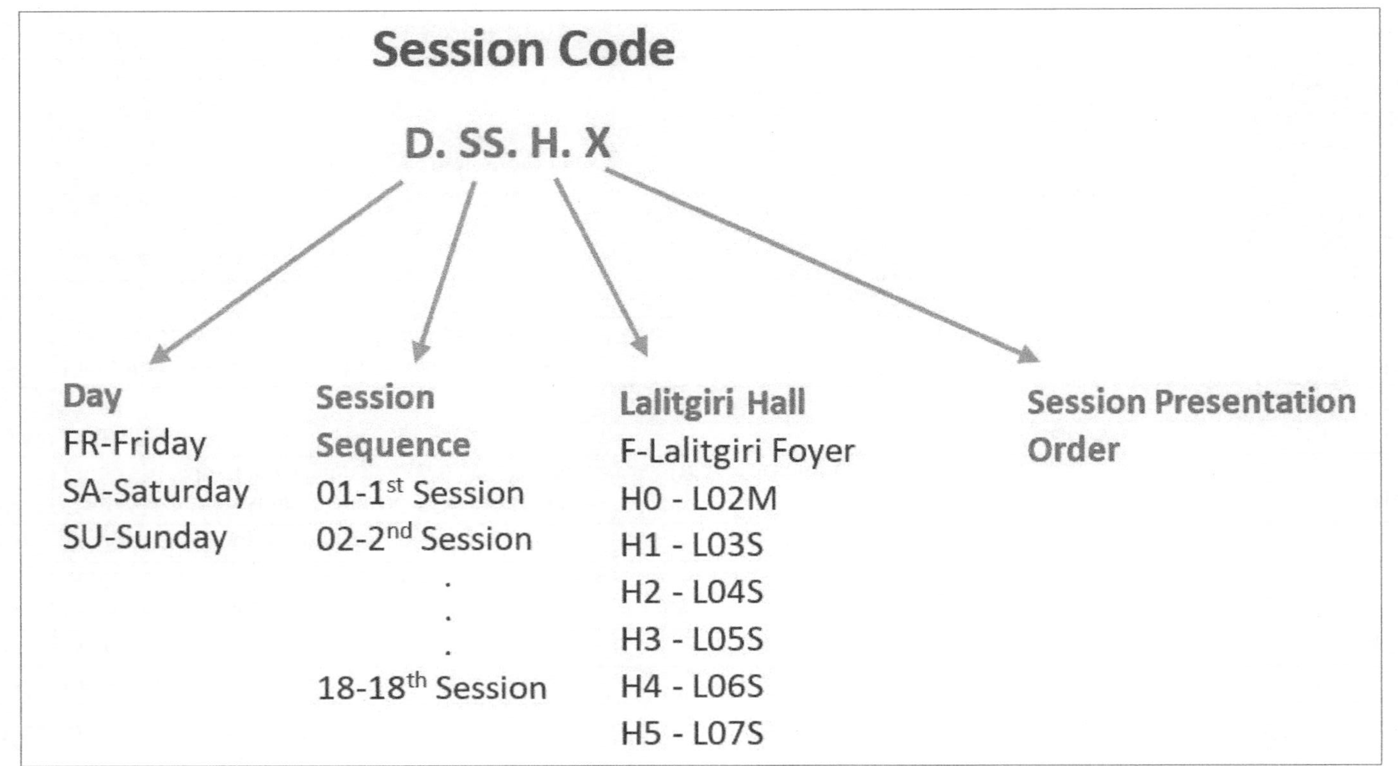

Conference Day 1: 23 May 2025

<table>
<tr><td>08:00</td><td colspan="5">Registration</td><td>H1</td></tr>
<tr><td>09:00</td><td colspan="5">Inaugural Session</td><td>H0</td></tr>
<tr><td>09:20</td><td colspan="5">Keynote Address 1: Prof. Siva Kumar, McMaster University, Canada
Topic: Recent Progress in Optical Communications</td><td>H0</td></tr>
<tr><td>10:10</td><td colspan="5">Tea Break + Group Photo Session</td><td>F</td></tr>
<tr><td>10:30</td><td colspan="5">Conference Workshop: Avenues of Research in 6G
Prof. P. K. Sahu, Dr. Debalina Ghosh, Dr. D. P. Acharya & Dr. S. Borkotoky</td><td>H0</td></tr>
<tr><td>12:30 to 13:30</td><td colspan="5">Lunch Break</td><td>F</td></tr>
<tr><td>13:30</td><td colspan="5">Keynote Address 2: Prof. Bratin Ghosh
Topic: Full-Wave Analysis and Design of Dielectric Resonator Antennas</td><td>H0</td></tr>
<tr><td rowspan="2">14:20 to 16:40</td><td colspan="6">Contributory Sessions</td></tr>
<tr><td>Session – FR.01.H1.X
Electromagnetic Modeling of Devices and Systems
Chairs:
Dr. S. Chander & Prof. Rajan Jha</td><td>Session – FR.02.H2.X
Propagation and Channel Modeling
Chair:
Prof. P. R. Sahu</td><td>Session – FR.03.H3.X
Optical Sensors and Detectors
Chairs:
Dr. Rajesh Kumar & Dr. Sourabh Pandey</td><td>Session – FR.04.H4.X
Advancements using AI
Chairs:
Dr. D. P. Dogra & Dr. Abhik Jana</td><td colspan="2">Session – FR.05.H5.X
Bioeffects and Medical Applications
Chairs:
Dr. H. Padole & Dr. Nitya Tiwari</td></tr>
<tr><td>16:40</td><td colspan="5">Keynote Address 3: Prof. Senthil M Ganapathy
Topic: Emerging Pathways in Biomedical Spectroscopy</td><td>Hall 0</td></tr>
<tr><td>17:30</td><td colspan="5">Tea Break</td><td>Foyer</td></tr>
<tr><td>17:40 to 18:40</td><td colspan="5">Industry Talks: Accelerating Innovations with AI
Dr. Pethuru Raj & Mr. Kamleshwar Rajender</td><td>Hall 0</td></tr>
</table>

Conference Day 2: 24 May 2025

9:15 to 11:00	**Contributory Sessions**			
	Session – SA.06.H1.X **Novel Antenna Designs** ***Chair:*** Prof. A. R. Harish	Session – SA.07.H2.X **AI in Communication System Design** ***Chairs:*** Prof. Mihir N. Mohanty & Soumya P. Das	Session – SA.08.H3.X **Optimization for Advanced Systems** ***Chairs:*** Dr. Sanjay Kumar Gowre	Session - SA.09.H4.X **Advancements in Healthcare using AI** ***Chairs:*** Dr. Veerendra D. & Dr. Satyasai J. Nanda
11:00	Tea Break + Group Photo Session			Foyer
11:20	**Keynote Address 4:** Prof. Vikram Gadre ***Topic:*** Looking at Two Domains at Once in System Theory and Differential Equations – Sharing Some of Our Recent Insights			Hall 0
12:10	**Invited Talk 1:** Dr. Ajit Jha ***Topic:*** The Convergence of mmWave Radar and Photonics for Next-Generation Sensing	Hall 1	**Invited Talk 2:** Dr. Varun Kumar ***Topic:*** Digital Holography and Its Applications in Engineering and Biosciences	Hall 2
13:00 to 14:00	Lunch Break			Foyer
14:00 to 15:45	**Contributory Sessions**			
	Session – SA.10.H1.X **Microwave and mmWave Sensors in Advanced Applications** ***Chair:*** Dr. Varun Kumar & Dr. Atri Mukhopadhyay	Session – SA.11.H2.X **Communication System Design and Optimization** ***Chair:*** Prof. P. Chakrabarti	Session – SA.12.H3.X **Automotive Sensing and Tracking** ***Chair:*** Dr. Rakesh Balabantaray	Session – SA.13.H4.X **Communication and Sensing on IOT Platforms** ***Chair:*** Dr. Siddhartha Borkotoky
15:45	**Keynote Address 5:** Prof. A. R. Harish ***Topic:*** Enhancing Antenna Performance with Metamaterials			Hall 0

16:40	**Invited Talk 3:** Dr. Amit Agrawal ***Topic:*** Integrated high-efficiency Metasurface Optics for Quantum Applications	Hall 1	**Networking Session**	Hall 2
17:30	Tea Break + Group Photo Session			Foyer
17:40	**Keynote Address 6:** Prof. P. Chakrabarti ***Topic:*** Terahertz Technology and its Potential Applications			Hall 0
18:30	Award Ceremony			Hall 0
19:00	Banquet Dinner			Foyer

Conference Day 3: 25 May 2025

9:00	**Keynote Address 7:** Prof. B. Sundarajan ***Topic:*** Function-Correcting Codes (FCC): A New Framework for Error Control Coding (ECC)				Hall 0
10:00	Tea Break				Foyer
10:15 to 12:30	Contributory Sessions				
	Session – SU.14.H1.X **Metasurfaces for Field Shaping and Synthesis** ***Chairs:*** Prof. S. Mahapatra & Dr. Srinivasarao C.	Session – SU.15.H2.X **Data Processing and AI for Imaging Applications** ***Chairs:*** Prof. A. K. Ojha & Dr. E. Sivasankar	Session – SU.16.H3.X **Advances in Optical Communication** ***Chairs:*** Dr. Bijayananda Patnaik & Dr. Ravi K. Maddila	Session – SU.17.H4.X **Sensing for Human Machine Interface** ***Chairs:*** Prof. B. Sundarajan	Session – SU.18.H5.X **Biomedical Imaging and Sensors** ***Chairs:*** Dr. N. B. Puhan & Dr. R P Padhy
12:30	Valedictory Session				Hall 1
12:45 to 13:30	Concluding Lunch				Foyer
13:30 Onward	Sightseeing Opportunities (Pre-booking)				

KEYNOTE ADDRESSES

Recent Progress in Optical Communications

Shiva Kumar

Professor, Electrical & Computer Engineering,
McMaster University, Hamilton, Ontario, Canada
E-mail: skumar@mcmaster.ca

Abstract: This Various options for the mitigation of linear and nonlinear distortions of fiber optic communication system in the optical domain are reviewed. Optical back propagation (OBP) using bidirectional Raman pumping is found to be an ideal candidate for the mitigation of nonlinear impairments in fiber optic networks. The possibility of implementing a reservoir computing (RC) based equalizer in the optical domain for the mitigation of impairments of a dispersion managed fiber optic system is investigated.

Shiva Kumar received the MS (1990) and Ph.D. degrees (1994) in electrical communication engineering from Indian Institute of Science, India, and a Ph.D. degree (1997) from Osaka University, Japan. He was a Humboldt fellow at University of Jena, Germany, supported by Alexander von Humboldt Foundation from 1997-98. He worked at Corning Incorporated, New York as a Senior Research Scientist (1998-2001), and then as a Project Leader (2001-2003). He is currently a Professor at McMaster University, Canada. He is also a visiting professor at Tohoku University, Japan from Sept. 2016 to June 2017 and Sept. 2023 to June 2024. Dr. Kumar is a Fellow of Optical Society of America (now Optica) since 2019. He has authored a textbook on fiber optic communications (2014), edited a book on nonlinear fiber optics (2011) and has published more than100 papers in refereed journals, 54 conference presentations, and 8 book chapters. He has delivered about 40 invited talks in international conferences/workshops. He was awarded 'R&D 100 award' for the product design 'Metrocor' by R&D Magazine, 'circle of excellence award' from the magazine Photonics spectra, and 'outstanding individual contributor award' by Corning Inc. He holds 10 US patents in the field of fiber optics and optoelectronics.

Full-Wave Analysis and Design of Dielectric Resonator Antennas

Bratin Ghosh

Professor, Department of E & ECE, IIT Kharagpur, India
E-mail: bghosh@ece.iitkgp.ac.in

Abstract: An efficient full-wave Green's function technique is presented for the analysis of the multilayer dielectric resonator antenna (DRA) structure with an arbitrary number of layers. Three planar/non-planar feed configurations are investigated, viz. the microstrip aperturecoupled feed, the probe-coupled feed and the rectangular waveguide fed multilayer DRA. The modes of the multilayer DRA can be identified from the analysis which can be used to explain the broadband nature of the coupling. The layer permittivities and thicknesses are optimized for broadband operation of the antenna structures. The mode-excitation for the antenna structures is thoroughly characterized using the Green's function approach, providing further insight into the radiation mechanism and bandwidth extension for such structures. Frequency tuning of the antenna structures is also demonstrated by exciting the antenna at a higher order mode, maintaining the broadband characteristics.

Bratin Ghosh received his Bachelors in Electronics and Telecommunication Engineering in 1990 from Jadavpur University, Kolkata, Master of Technology in Electronics and Electrical Communication Engineering with specialization in Microwave Engineering in 1994 from the Indian Institute of Technology, Kharagpur and Ph.D. in Applied Electromagnetics in 2002 from the University of Manitoba, Canada. He is currently Professor in the Department of Electronics and Electrical Communication Engineering, Indian Institute of Technology, Kharagpur. He is a TPC member and an invited/keynote speaker and chaired sessions in many national and international conferences and seminars. He is an Associate Editor for the IEEE Transactions on Antennas and Propagation and the IEEE Antennas and Wireless Propagation Letters. He is also in the review and editorial boards of many international journals. He has been the recipient of many federal project grants in addition to research grants from the industry on the design of efficient antennas and guided systems. He has also organized many short term lectures and workshops in the field of applied electromagnetics that have been well attended by participants from both the academia and industry. He had been the recipient of the National Talent Search Scholarship, the University of Manitoba Graduate Fellowship and is Senior Member, IEEE. He has also been the recipient of many national awards, travel grants and best student paper awards. His areas of research interest are full-wave Green's function technique for the analysis of cylindrical and spherical multilayer structures, singularity removal techniques, efficient evaluation of Sommerfeld integral tails, leaky and creeping waves in cylindrical structures and anisotropic media. He is also actively involved in the design and analysis of efficient horn antenna feeds, metamaterials, phased arrays, dielectric resonator antennas, antenna miniaturization and numerical techniques. He has also edited a monograph on the full-wave analysis of guided wave structures and components.

Emerging Pathways in Biomedical Spectroscopy

Senthil Murugan Ganapathy

Professor, University of Southhampton, Southampton, United Kingdom
E-mail: smg@orc.soton.ac.uk

Abstract: The ability to detect and quantify low concentrations of analytes in a simple, low-cost, and rapid manner using minimal sample volumes is essential for biomedical diagnostics, enabling early detection and diagnosis. Vibrational spectroscopy is a label-free technique capable of identifying and quantifying various biomarkers. The focus of this keynote talk will be on vibrational spectroscopy, including infrared (IR) and Raman modalities, which allow rapid, label-free molecular characterisation. To address longstanding challenges such as bulky instrumentation and low detection sensitivity, we have developed a silicon-based attenuated total reflectance (ATR) platform enhanced with Metasurfaces, significantly improving signal intensity and biomolecular detection. Additionally, an adaptive atmospheric correction algorithm ensures spectral accuracy without the need for nitrogen purging, while the integration of IR and Raman spectral data with machine learning further enhances diagnostic precision. These advancements have significant biomedical implications, including the early detection of neonatal respiratory distress syndrome (nRDS) and the quantification of cancer biomarkers from blood samples. By combining miniaturisation, portability, and multimodal capabilities, this work paves the way for point-of-care solutions, particularly in resource-limited settings.

Senthil Murugan Ganapathy is a Professor and Head of the Integrated Photonic Devices Group, as well as Deputy Head of School (Education) at the Optoelectronics Research Centre, University of Southampton, United Kingdom. He received a PhD degree in Photonic Materials in 2001 from the Indian Institute of Science, Bangalore. From 2001 to 2005, he worked as a postdoctoral fellow at the University of Bordeaux, France, and the Toyota Technological Institute, Japan. He joined the Optoelectronics Research Centre at the University of Southampton in 2005.

His expertise and research interests span from photonic materials to photonic systems, with a current focus on mid-IR/high-contrast materials and devices for biomedical sensing, on-chip spectroscopy, on-chip nanoscopy, environmental monitoring, green energy generation, and optical communication applications. He has published more than 300 journal and conference papers, including over 40 invited/keynote talks at major international conferences, and is the editor of the book *Photonic Glasses and Glass-Ceramics.*

He is a Fellow of the Higher Education Academy of the UK and an Editor of the *Journal of Materials Science: Materials in Electronics*, published by Springer Nature.

Looking at Two Domains at Once in System Theory and Differential Equations - Sharing Some of Our Recent Insights

Vikram M Gadre

Professor, Department of Electrical Engineering, IIT Bombay, India
E-mail: vmgadre@ee.iitb.ac.in

Abstract: In communication, signal processing, image processing or system theory, it is tempting to see if we can get the best of both the worlds - the natural domain of representation and the transform domain or representation, in terms of an alternate basis. The whole world of multiresolution analysis was born of the desire to look, at once, in the time domain and in the Fourier domain. Broadening this endeavour, we have been delving into other perspectives, where we can look at entities in two domains at once. In this talk, some recent insights which we have, in the context of linear and nonlinear systems; as also in the context of 'self-dual' differential equations; which seem to open up a not-so-well-known connection between whole classes of known concepts and entities in communication, signal processing, image processing and system theory are elaborated upon. In the first part of the talk, some self-dual class of differential equations and the concomitant surprises, are brought out. In the second part, we look at an alternate representation of linear and non-linear systems by a dual representation and bring out possibilities for their application in system realization. We also point to some less known extensions of the uncertainty principle, a well understood concept in its elemental form in communication; noting that it points to the limitations of dealing with two domains at once. We also point to the applications and implications of these insights in communication and signal processing.

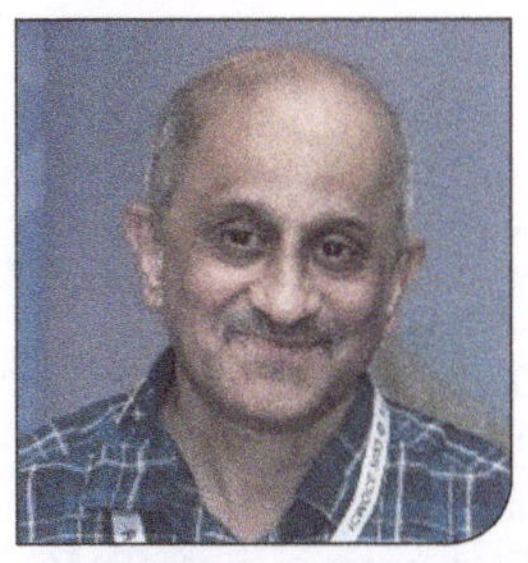

Vikram M. Gadre is a Professor in the Department of Electrical Engineering, IIT Bombay. He has 30 years of experience as a faculty member at IIT Bombay - with his involvement in teaching, research and administration. His principal area of research is in wavelets, multirate and multiresolution signal processing, filter banks and time-frequency principles: including both theory and applications. He has received the INSA Teachers' Award in Dec 2017, the Award for Excellence in Teaching at IIT Bombay in 1999, 2004, 2009, 2014 and 2024. He has also received the S P Sukhatme Award for Excellence in Teaching at IIT Bombay, in 2018. He has served as a member of the IIT Council, nominated by the Visitor (President of India) from July 2017 to February 2021.

Enhancing Antenna Performance with Metamaterials

A R Harish

Professor, Department of Electrical Engineering, IIT Kanpur, India
E-mail: arh@iitk.ac.in

Abstract: Antenna, an integral part of a wireless system, is evolving towards achieving better performance while shrinking in size. One of the more recent strategies to achieve this has been to use metamaterials – "artificially engineered materials exhibiting electromagnetic properties not found in nature". This talk is an attempt to understand "meta" in "metamaterial", explore the tools and techniques used to characterize the metamaterials, and apply this knowledge to improve the performance of antennas.

A.R. Harish is a professor in the Department of Electrical Engineering, Indian Institute of Technology Kanpur, Kanpur. He is part of a team that works on antennas, electromagnetics, meta-surfaces, RFID systems, EMI/EMC, etc. He enjoys teaching electromagnetics, antennas, and their applications.

Terahertz Technology and Its Potential Applications

Parthasarathi Chakrabarti

Former Professor, Department of Electronics Engineering, IIT (BHU), India
E-mail: pchakrabarti.ece@iitbhu.ac.in

Abstract: Terahertz (THz) waves also known as T-waves or T-rays are electromagnetic waves that encompass the frequency band lying between millimeter wave and infrared region of the electromagnetic spectrum (0.3 THz to 30 THz). More often the band from 0.1 THz to 10 THz is considered as the THz band as prescribed by ITU for 6G and beyond. The potential of Terahertz band mostly remained unexplored and unexploited until in recent times when THz waves turn out to be very promising for several telecommunication and non-telecommunication applications in both military and civil situations. In the field of telecommunication, THz technology will emerge as a major game changer in wireless data transfer leading to ultra high-speed internet (data rate~Tbps) with extremely low latency (~µs), seamless connectivity and robust data security enabling superior implementation of virtual, augmented and extended reality and finally leading to truly immersive experience by creating multiple sensorial media (Mulsemedia) environment. The non-ionizing nature of THz waves make them suitable for several non-telecommunication applications including non-destructive fault testing in advanced semiconductor packaging, ceramic and other composite material structures, high-resolution imaging in healthcare, smart security system and quality control of food and pharmaceutical products. The present talk makes an attempt to capture the advances in THz technology for implementation of THz communication, sensing and imaging. The talk also highlights the contribution of the author's research group in this field and outlines the challenges for successful exploitation of the full potential of this band.

Parthasarathi Chakrabarti is a Former Professor in the Department of Electronics Engineering, IIT(BHU). He served as the Director of MNNIT Allahabad during 2011-2016 and the Director of the Indian Institute of Engineering Science & Technology (IIEST), Shibpur during 2018-2024. He has made a significant contribution to experimental and theoretical research in the field of Microelectronics and Photonics. He has been the Coordinator (2005-2011) of the Centre for Research in Microelectronics, IIT(BHU), and the founder Coordinator of the MEMS Centre at IIT(BHU). Twenty-three (23) students including one international student under the ICCR Exchange Program have been awarded Ph.D. degrees under his supervision. He has successfully completed several R & D projects sponsored by the funding agencies like DST, CSIR, DRDO, etc. Prof. Chakrabarti has published over 350 research papers in leading technical journals and in Conference Proceedings. He has authored 4 textbooks and edited 6 Proceedings of International Conferences. He holds 4 patents. He is a recipient of INSA Visiting Fellowship and the EPSRC (UK) Visiting Fellowship for carrying collaborative research with Lancaster University, UK. He is a Fellow of the Institution of Engineers and a Senior Member of IEEE, USA. He is a recipient of the MRSI Distinguished Lecturer-ship Award 2019-20.

Function-Correcting Codes (FCC): A New Framework for Error Control Coding (ECC)

B. Sundar Rajan

Professor, Department of Electrical Communication Engineering, Indian Institute of Science, Bangalore, India
E-mail: bsrajan@iisc.ac.in

Abstract: In the classical Error Control Coding (ECC) problem k message bits are encoded into n transmitted bits and the receiver wants to recover all the k message bits even if there are up to t errors in the n transmitted bits. Consider the scenario where only a certain attribute of the k message bits is of interest to the receiver and not the entire k message bits and this attribute is the result of a certain function of the k message bits, say f(k message bits). Now the receiver wants the function value to be error-free even if t transmitted bits are erroneous. That is, it is not necessary for the receiver to recover all the k message bits correctly since it is more interested in the function values of the k message bits. Codes designed to achieve this additional protection to the function values and not the entire message bits are called Function-Correcting Codes (FCCs). Note that for the special case of the function f(.) being a bijection the theory of FCC coincides with the theory of ECC. That is, ECCs are a special case of FCCs. Recently it has been shown that FCCs can be designed with a small number of redundant bits. In other words, the length n of FCCs can be much smaller compared to the length of the ECCs. This amounts to large savings in bandwidth and the amount of savings in bandwidth depends on the function f(.). In this talk we will discuss all the results on FCCs known in the literature for various classes of functions. It turns out that several well-known bounds for ECCs like Singleton bound, Plotkin bound, and Gilbert-Varshamov bound can be generalized to FCCs. The theory of FCCs is in its infant stage and can be considered as a part of semantic communication. New results on bounds and FCCs for the special case of the function f(.) being a linear function will be presented along with highlighting several open problems.

B. Sundar Rajan is a Life Fellow of IEEE and was a member of Fellow Evaluation Standing Committee of IEEE Communications Society (2019-2021). He was an Associate Editor of IEEE Transactions on Information Theory (2008-2011 and 2013-2015), an Editor of IEEE Transactions on Wireless Communications (2007-2011) and an Editor of IEEE Wireless Communications Letters (2012-2015). He served as a Technical Program Co-Chair of the IEEE Information Theory Workshop (ITW'02), held in Bangalore, in 2002. He is a fellow of the Indian National Academy of Engineering, the Indian National Science Academy, the Indian Academy of Sciences, and the National Academy of Sciences, India. He is a recipient of Prof. Rustum Choksi Award by Indian Institute of Science for Excellence in Research in Engineering for the year 2009, the IETE Pune Centre's S.V.C Aiya Award for Telecom Education in 2004, and Best Academic Paper Award at the IEEE WCNC 2011. Prof. Rajan received his B.Sc. degree in mathematics from Madras University, B.Tech. degree in electronics from Madras Institute of Technology, India, and M.Tech. In addition, Ph.D. degrees in electrical engineering from IIT Kanpur. He was a faculty member at the Department of Electrical Engineering, IIT Delhi, from 1990 to 1997. He has been a Professor with the Department of Electrical Communication Engineering at Indian Institute of Science, Bangalore, since 1998. He is a J.C. National Fellow (2016-2025). His current research interests include Coded Caching, Private Information Retrieval (PIR), Coding for Distributed Computing, Semantic Communication, Over-The-Air (OTA) computing, and Coding for MIMO and multi-user communication.

INVITED TALKS

The Convergence of mmWave Radar and Photonics for Next-Generation Sensing

Ajit Jha

Associate Professor, Department of Engineering Sciences,
University of Agder, Norway
E-mail: ajit.jha@uia.no

Abstract: Millimeter-wave (mmWave) radar and laser-based photonic systems have each emerged as powerful sensing modalities. While former offers robustness and depth penetration, latter deliver precision and high-resolution optical capabilities. Traditionally treated as distinct technologies, their convergence opens new frontiers in sensing by combining the strengths of electromagnetic and optical domains. This talk explores how mmWave radar and laser-based photonics can be co-designed and integrated to overcome the limitations of each—enhancing flexibility, resolution, and adaptability across a range of applications including – autonomous vehicle, condition-based asset integrity, predictive maintenance, and biomedical diagnostics. Further, we will discuss emerging architectures that use lasers for mmWave signal generation, laser-radar fusion for enhanced functionality. Finally, we will examine key research challenges, such as integration, and cross-domain signal processing, and present a vision for unified sensing platforms that leverage both radar and laser technologies for intelligent and adaptive sensing.

Ajit Jha is an Associate Professor with Department of Engineering Sciences at University of Agder, Norway. He is engaged in cutting-edge research in Photonic Sensing, IoT, Machine Learning, Bio-photonics, and Sensor Fusion. His work spans transformative applications like laser vibrometer, optical feedback to Autonomous Systems, Precision Agriculture, and Structural Health Monitoring. With a rich academic and industrial background, Ajit holds a joint PhD in Photonics Engineering Mechanical Engineering from Technical University of Catalunya (UPC), Spain, and Karlsruhe Institute of Technology (KIT), Germany, under Erasmus Mundus Joint Doctorate EUROPHOTONICS program funded by the EU. Before his PhD, Ajit worked as a researcher at Fraunhofer Heinrich Hertz Institute, Berlin on optical wireless communication. In industry, Ajit worked on technologies related to in Autonomous Vehicles, ADAS, Surround View Systems, technologies. He has published more than 60 manuscripts in various journals including IEEE Photonic Technology Letters, Optic Express, Applied Optics, IEEE Transactions on Geoscience and Remote Sensing. He is also committed to contributing to the academic community, he actively organizes IEEE conferences and serves on technical program committees, shaping the future of sensing, and photonics.

Digital Holography and Its Applications in Engineering and Biosciences

Varun Kumar

Associate Professor, Institute of Science and Technology, Shimane University, Japan
E-mail: varunkumar@ecs.shimane-u.ac.jp

Abstract: Holography is a lens less imaging technique that facilitates the recording and precise reconstruction of object wavefront (3D information) of an object under investigation. In conventional holography, hologram as a micro-interference pattern between the object wavefront and reference wave is recorded on high resolution photographic plate. Digital holography (DH) replaces the photographic plate with opto-electronics recording devices like chargecoupled device (CCD) or complementary metal oxide semiconductor (CMOS) camera for the recording of holograms. DH provides pixel by pixel quantitative information of the complex amplitude (amplitude and phase) of the object wavefront from a single digital hologram by applying an appropriate numerical reconstruction method. Advancement in computer technology, high resolution, high dynamic range, fast recording devices and coherent light sources has made it possible to use digital holographic systems for numerous industrial/engineering and biosciences applications. In this paper, DH is presented for its emerging industrial applications ranging from non-destructive, noninvasive full field temperature distribution measurement, heat transfer from heated objects, temperature field inside the flames, magnetic promotion of combustion in flames, testing of micro-optics, 3D imaging of biological cells (RBCs, WBCs, bacteria etc.) to shrinkage behavior of ultraviolet (UV)- curable adhesive.

Varun Kumar is an Associate Professor at Institute of Science and Technology in Shimane University, Japan. He received his MTech. degree in Instrument Technology and Ph.D. in Optical Engineering both from IIT Delhi in 2010 and 2016, respectively. After completing his Ph.D., he has been a Research Fellow at NTU Singapore, and Assistant Professor at NIT Hamirpur, H.P, India. His research interest includes digital holography, holographic microscopy, interferometry, optical metrology, development of novel industrial and biomedical applications of digital holography and laser-based instrumentation. He has more than 45 publications in international journals, conference proceedings and book chapters. He is a member of SPIE and OSA.

Integrated high-efficiency Metasurface Optics for Quantum Applications

Amit Agrawal

Associate Professor, Department of Engineering, University of Cambridge, UK
E-mail: aka59@cam.ac.uk

Amit Agrawal is an Associate Professor of Optical Engineering at the University of Cambridge in the UK. He received Bachelors in Electronics and Telecommunications Engineering from India and, M.S. and Ph.D. in Electrical Engineering from the University of Utah. Following that, he did his postdoc at UMD/NIST, where he worked in the areas of Plasmonic and metamaterials. He then joined the faculty of Syracuse University for three years and subsequently served on Staff Scientist roles at NIST and the US Army Research Lab. At Cambridge, his group works in the areas of ultrafast optics, nonlinear optics, nanofabrication and integrated photonics. His research work in the last few years has primarily focused on developing Metasurfaces and integrated nanophotonic devices operating from the ultraviolet to the near-infrared to scale and/or miniaturize quantum systems based on atoms and ions for applications in quantum sensing and computing.

INDUSTRY TALKS

Accelerate AI Innovation with the Dell AI Factory

Kamleshwar Rajender

AI Specialist, AI APJ Team, Dell Technologies, India
E-mail: Kamleshwar.Rajender@dell.com

Abstract: AI will be the defining technology of the years to come, and we want to help organizations set themselves on the path to success. The AI revolution is here, and it's transforming business at an unprecedented pace. Success in the evolving global economy will depend on how quickly and effectively organizations can harness the potential of AI. To stay relevant, accelerating toward an AI future is no longer a choice; it's imperative. However, navigating the AI opportunity poses challenges—from addressing skills gaps to data sovereignty and the costs of supporting and scaling AI across an enterprise. The Dell AI Factory addresses these most pressing challenges with a collection of AI technologies, an open ecosystem of partners, validated and integrated solutions and expert services to help you achieve AI outcomes faster and with the next generation infrastructure to sustain AI workloads. The Dell AI Factory is more than just a set of services or products; it's our vision to help organizations harness the full potential of AI. With our broad AI portfolio and leading infrastructure, an open ecosystem and services offerings, we are helping organizations around the world build their own AI factories to transform and innovate.

Kamlesh is a roll up your sleeves Technology professional with 15+ years in Technology Sales and Solution Selling role with customers across various technology segments. Kamlesh serves as the AI Specialist for the Dell Technologies, AI APJ Team. Through industry specific AI Go-To-Market and Sales strategies, Kamlesh has been spearheading AI based Digital Transformations for large enterprises across various industries, enabling them in their AI Journey. Before joining Dell Technologies in 2024, Kamlesh spent more than a decade in Selling AI Infra, cloud, HPC, AI Services and Solutions with Oracle and Altair. In this capacity, he served as a trusted advisor to a multitude of clients in the government, health care, life sciences, manufacturing, BFSI, retail, academic, and geospatial sectors etc. Kamlesh's other professional experiences include cloud computing programs, AI Simulation software's etc. He holds an Engineering degree from Visvesvaraya technological university, MS in Automotive engineering from Coventry University MBA from Liverpool Business School, UK and PGDM from IMT, Ghaziabad.

Illustrating Edge AI 2.0 Techniques and Tools for Real-world Digital Transformation

Pethuru Raj

Vice President and Chief Architect, Edge AI Division, Reliance Jio Platforms Ltd, India
E-mail: Pethuru.Chelliah@ril.com

Abstract: With the astounding growth in the artificial intelligence (AI) technology ecosystem, a variety of business processes across industry verticals are being attempted to be automated and accelerated. Today we have a bevy of pioneering AI algorithms, models and agents empowering business behemoths and start-ups to be right and relevant to their customers and consumers. With the ready availability of big data and greater computational power through AI-centric processing units, AI-based data analytics brings forth predictive, prescriptive and personalized insights in time. The knowledge discovered gets disseminated to appropriate systems and devices to exhibit intelligent behaviour in their assignments and obligations. In this talk, I will focus on the following topics.

1. Edge Computing for enabling On-Device Data Processing
2. Edge Analytics for extracting Real-time Insights out of Edge Data
3. Edge Cloud Formation using lightweight versions of the Kubernetes Platform
4. AI Model Optimization Techniques to run AI models on Edge Devices
5. Edge AI for Real-time Intelligent Services and Applications
6. Edge AI 2.0 for running generative AI models on edge devices

Pethuru Raj, PhD is a chief architect at the Edge AI division of Reliance Jio Platforms Ltd. (JPL) Bangalore. Previously. Worked in IBM Global Cloud Centre of Excellence (CoE), Wipro consulting services (WCS), and Robert Bosch Corporate Research (CR). He had gained over 24 years of IT industry experience and nine years of research experience. He has finished the CSIR-sponsored PhD at Anna University, Chennai. He continued with the UGC-sponsored postdoctoral research in the Department of Computer Science and Automation, Indian Institute of Science (IISc), Bangalore. After that, he got two international research fellowships (JSPS and JST) to work as a research scientist for 3.5 years in two leading Japanese universities. He focuses on some of the emerging technologies, such as Multimodal Generative and Agentic AI, AI Model Optimization Techniques, the Internet of Things Use Cases, Cloud-native and Edge Computing Paradigms, Reliability Engineering, 6G Communication and Blockchain Technologies, Quantum-safe Cryptography Algorithms.

CONFERENCE WORKSHOP

Avenues of Research in 6G

Abstract: The rapid evolution of wireless communication systems has ushered in a new era of technological convergence, positioning Sixth Generation (6G) networks as the cornerstone of future connectivity. This workshop, titled *"Avenues of Research in 6G,"* will explore the foundational directions, disruptive technologies, and emerging challenges shaping the global 6G research landscape. Topics will include terahertz communication, intelligent reflecting surfaces (IRS), ultra-massive MIMO, integrated sensing and communication (ISAC), and free space optical communications. The workshop is aimed at fostering interdisciplinary research and will feature four expert talks on Reconfigurable and Programmable Metamaterials for 6G, Free Space Optical Communication for Advancing 6G Connectivity and Intelligent IoT in 6G. Participants will gain insights into the theoretical underpinnings, practical considerations, and policy frameworks necessary to drive innovation in 6G. This platform invites academia, industry, and government stakeholders to co-create the roadmap for next-generation wireless systems.

Workshop Talk 1:
6G Communications: From Bits to Semantics

Siddhartha S. Borkotoky

Assistant Professor, School of Electrical and Computer Sciences,
IIT Bhubaneswar, India
E-mail: borkotoky@iitbbs.ac.in

Abstract: We explore the topic of reliable information delivery in next-generation wireless networks. We begin with an overview of current trends in modern error-correction coding techniques aimed at ensuring bit-level accuracy in 6G systems. Next, we contrast the traditional view of data as a collection of bits with the perspective that considers data as the conveyor of meaning. This shift leads us to the concept of semantic communication, where the goal is to convey the meaning of information using semantic metrics and context-aware inference. We examine how semantic communication fits into the 6G landscape and the potential it holds for enabling future intelligent networks.

Siddhartha Borkotoky is an Assistant Professor with the School of Electrical and Computer Sciences at the Indian Institute of Technology Bhubaneswar. He received the PhD degree in Electrical Engineering from Clemson University, USA, and subsequently worked as a postdoctoral researcher with Clemson University and as a senior researcher with Lakeside Labs GmbH at Klagenfurt, Austria before joining in his current position. His research interests include wireless sensor networks, application-layer coding, and cooperative communications.

Workshop Talk 2: Reconfigurable and Programmable Metamaterials in 6G Communication Systems

DEBALINA GHOSH

Associate Professor, School of Electrical and Computer Sciences,
IIT Bhubaneswar, India
E-mail: deghosh@iitbbs.ac.in

Abstract: Reconfigurable and programmable metamaterials are emerging as transformative technologies poised to redefine the physical layer of 6G communication systems. These engineered surfaces, capable of dynamically manipulating electromagnetic waves in space, time, and frequency, offer unprecedented control over signal propagation, enabling functionalities such as beam steering, adaptive coverage, interference suppression, and physical-layer security. This talk will delve into the design principles behind metamaterials and highlights the control strategies behind programmable metamaterials such as intelligent reflecting surfaces (IRS). This session will illuminate the practical challenges and research opportunities in realizing truly reconfigurable electromagnetic environments for next-generation wireless communication.

Debalina Ghosh is an Associate Professor in the School of Electrical and Computer Sciences at IIT Bhubaneswar, India and is a key member of the Microwave and Optical Engineering (MOE) Research Group, which focuses on multi-disciplinary areas of microwave and optical engineering. She has 25 years of experience successfully contributing to research in the domain of applied electromagnetics including the design and optimization of antennas and antenna arrays, microwave absorbers and polarizers, reconfigurable intelligent surfaces (RIS) and radar systems. In 2016, Dr. Ghosh was elevated to the grade of Senior Member of the IEEE, recognizing her professional accomplishments. She has supervised several Ph.D. students, with completed theses on diverse topics such as aperiodic antenna array optimization, Metasurface antennas, microwave sensing for bone health evaluation, and human vital signs detection through obstructive environments.

She is also involved in a number of research projects funded by ANRF, MEITY and DRDO, Govt. of India with applications in remote sensing and healthcare, in personal and satellite communications and in radar design. Previously she has worked as a Senior Antenna Design Engineer in California, USA. She received the BE from Jadavpur University, and the MS and PhD from Syracuse University, New York, USA. Additionally, Dr. Ghosh has played a significant role in organizing various academic events, serving as the General Chair for three editions of the International Conference on Microwave, Optical and Communication Engineering (ICMOCE) organized at IIT Bhubaneswar.

Workshop Talk 3: Free Space Optical Communication for Advancing 6G Connectivity

Prasant Kumar Sahu

Professor, School of Electrical and Computer Sciences, IIT Bhubaneswar, India
E-mail: pks@iitbbs.ac.in

Abstract: Free Space Optical (FSO) communication systems have recently attracted renewed attention as a promising wireless interconnect solution for the 5G and beyond communication as well as for 6G. FSO supports high capacity, low latency, energy efficiency, dense connectivity, and strong security against interception makes them well suited to meet the demands of next-generation networks. In this presentation, the objective is to present the key challenges and opportunities associated with deploying FSO systems in 6G networks. FSO communication provides an excellent opportunity to develop ultrafast data links that can be applied in a variety of 6G applications, including heterogeneous networks with enormous connectivity and wireless backhauls for cellular systems.

Prasant Kumar Sahu currently working as Professor in the School of Electrical and Computer Sciences of IIT Bhubaneswar. He did his M. Tech in Electronics and Electrical Communication Engineering with Fiber Optics and Lightwave Engineering specialization form IIT Kharagpur and Ph. D. in Electronics and Electrical Communication Engineering from Indian Institute of Technology Kharagpur. He also worked in the University of Scoula Superiore Sant'Anna Italy as a visiting Scholar for a year in the field of signal processing in distributed Sensor design. Before joining IIT Bhubaneswar, he was working as a faculty in the ECE dept. of NIT Rourkela. His research interest includes Long-Haul Communication System Design, Signal Processing, Design and Development of Intelligent Sensors, Distributed Sensors, Optical Instrumentations, Speech Recognition, ICT, Healthcare. He has about 100+ Publications in Reputed International and National Journals and in Proceedings of Conferences. He has supervised a good number of Ph.D, M.Tech and B.Tech students. He is having 25 Years of Academic Experience. He has carried out a number of sponsored research projects. He has reviewed many peered reviewed journals like Optical Engineering, Sensor Letter, Measurement, Photonics Technology Letter, JLT, Measurement, Optical Communication etc. He is a Senior Member IEEE, Fellow IETE (I), Fellow IE (I) and Senior Member OSA etc.

Workshop Talk 4: Intelligent IoT in 6G

Debiprasad Priyabrata Acharya

Professor, Department of Electronics & Communication Engineering,
National Institute of Technology Rourkela, India
E-mail: dpacharya@nitrkl.ac.in

Abstract: Unprecedented data rates, ultra-low latency, and ubiquitous connectivity are the promise of sixth-generation (6G) networks. This next generation of wireless infrastructure is expected to serve as a catalyst for the Intelligent Internet of Things (IIoT), integrating advanced artificial intelligence (AI) with massively interconnected devices. In a 6G-enabled IIoT ecosystem, real-time data processing, autonomous decision-making, and edge intelligence become core features, enabling applications such as smart cities, autonomous vehicles, precision agriculture, and advanced healthcare. Key technologies such as terahertz communication, intelligent reflecting surfaces (IRS), distributed AI, and quantum communication are expected to underpin this paradigm. This talk explores the convergence of AI and IoT within the 6G framework, and outlines future directions for research and development in achieving a truly intelligent and interconnected world.

Debiprasad Priyabrata Acharya is Professor in the Department of Electronics & Communication Engineering at National Institute of Technology Rourkela, where he has been a faculty member since 2008. He also holds the position of Technical Coordinator at the Centre for IoT and Embedded Technologies, reflecting his leadership in advancing research and education in these critical fields. His expertise encompasses a wide range of areas, including the Internet of Things (IoT), Digital and Analog VLSI, Embedded Systems and Machine Intelligence. Prof. Acharya has made significant contributions to research, authoring over 30 peer-reviewed journal articles and more than 24 conference papers. His research interests are not only theoretical but also practical, as evidenced by his ongoing projects, such as the development of an embedded system for compressed air monitoring in a smelter plant and an intelligent leakage management in fly ash slurry lines. In addition to his research, Prof. Acharya has been actively involved in teaching and mentorship, supervising five completed PhD candidates and six ongoing, along with overseeing 65 completed M.Tech projects and many BTech projects. He teaches Digital VLSI Design, Embedded Computing Systems, and IoT, shaping the next generation of engineers in these vital areas. Prof. Acharya's academic foundation is robust, with a Master of Science in Physics with Electronics from the Indian Institute of Technology Kharagpur followed by Master of Engineering in Microelectronics from the Birla Institute of Technology and Science Pilani. He completed his Doctoral research in Electronics & Communication Engineering at National Institute of Technology Rourkela. His early industrial career started as a research engineer at the Centre for Development of Telematics (C-DOT), Bangalore where he contributed to the advancement of telecom technology in India. His professional journey includes significant industry experience, having worked as a Technical Consultant in the VLSI Business Unit at Wipro Technologies, Bangalore and Specialist in ASIC/FPGA at Tata Elxsi Ltd, Bangalore.

He has been consultant for providing technical solution to problems in different industries like JSPL, Sankalp Semiconductors, NALCO, Apstronics and many other industries. His academic and professional achievements are complemented by his commitment to innovation, as evidenced by his patents, one of which has been granted and another published. He has been a member of VLSI Society of India (VSI), a life member of Institution of Electronics and Telecommunication Engineers (IETE) and a Fellow, Institution of Engineers (IE), India. He is also closely involved in many social activities both as a team member and as leader at NIT Rourkela.

SCHEDULE OF CONTRIBUTORY SESSIONS

Conference Day 1: Friday, 23rd May 2025

Session – FR.01.H1.X

Electromagnetic Modeling of Devices and Systems
Chair: Dr. Subhash Chander (SSPL, DRDO) & Prof. Rajan Jha (IIT BBS)
Location: Hall 1 (L03S)
Time: 14:20 to 16:40

FR.01.H1.1	Paper ID 48: Design and Realization of Gysel Power Divider & Combiner for High Power Handling in C-Band	Manish Mishra *Bharat Electronics Limited*
FR.01.H1.2	Paper ID 259: Design and Simulation of C-Band 500 MW Relativistic Backward Wave Oscillator	Sanjay Gupta *MTRDC, DRDO*
FR.01.H1.3	Paper ID 262: Large signal GaN HEMT Model up to Ku Band for Switch Applications	Kirti Bansal *SSPL, DRDO*
FR.01.H1.4	Paper ID 329: Precision phase matched low Noise Figure high gain C Band RF Down Converter	Narendra S *Bharat Electronics Limited, Bangalore*
FR.01.H1.5	Paper ID 332: Graphene-Inspired THz Frequency-Reconfigurable Antennas for Biomedical Applications	Shweta Thomas *VIT, Vellore*
FR.01.H1.6	Paper ID 350: Examining the possibility of rGO@PANI with co-doped Mn_2O_3 and $CoFe_2O_4$ composites as a radar-absorbing material for the whole X and Ku bands	Parveen Kumar *NIT Kurukshetra*
FR.01.H1.7	Paper ID 495: High efficiency High Output and Wideband GaN HEMT Power Amplifier over entire S-Band	Pradeep Gorre *Vignan's Foundation for Science, Technology & Research*
FR.01.H1.8	Paper ID 512: Standing Wave Pattern and Frequency Measurement System using Capacitively Coupled Probes for Pedagogical Applications	Debapratim Ghosh *IIT Bhubaneswar*
FR.01.H1.9	Paper ID 564: Design of a Wideband Millimeter-Wave 5th Order Chebyshev CPW Bandpass Filter	Ashwini Sawant *IIT Guwahati*

Session – FR.02.H2.X

Propagation and Channel Modeling
Chair: Prof. P. R. Sahu (IIT BBS)
Location: Hall 2 (L04S)
Time: 14:20 to 16:20

FR.02.H2.1	Paper ID 165: Channel Estimation for Intelligent Reflecting Surfaces for MISO-OFDM System by Deep Learning	Subhra Sahu *NIT Rourkela*
FR.02.H2.2	Paper ID 183: Ensemble Learning-based Holistic Attack Classification for Enhanced Physical Layer Security	Swethambri M *Anna University*
FR.02.H2.3	Paper ID 208: Modulation Classification: Diverse Techniques and Constellation Model Training at Varying Resolution	Varun Kumar *IIT Vadodara*
FR.02.H2.4	Paper ID 215: Analytical ABEP and Diversity Order of IRS-Assisted Wireless Systems in Nakagami-m Fading	Sanjeet Bhagat *IIT Bhubaneswar*
FR.02.H2.5	Paper ID 251: A Novel Precoder for Peak-to-Average Power Ratio Reduction in OTFS Systems	Saif Khan Mohammed *IIT Delhi*
FR.02.H2.6	Paper ID 276: A Lightweight Video Streaming Framework for Low-End Systems Using FFMPEG and UDP for Real-Time Transmission	Neha Gupta *Bharat Electronics Limited*
FR.02.H2.7	Paper ID 348: A Comparative Analysis of Advanced Modulation and Coding Schemes for LEO Based Satellite Network	Deepika Manjunath *Bharat Electronics Limited*
FR.02.H2.8	Paper ID 353: Higher Order Time-Frequency Uncertainty Products and Self-Dual Equations	Vineet Ghule *IIT Bombay*

Session – FR.03.H3.X

Optical Sensors and Detectors
Chair: Dr. Rajesh Kumar (IIT Roorkee) & Dr. Sourabh Pandey (IIT Patna)
Location: Hall 3 (L05S)
Time: 14:20 to 16:40

FR.03.H3.1	Paper ID 88: Assessing Fiber Bragg Grating Sensors for Landslide Monitoring: Current Trends and Future Prospects	Ajish U *Amrita Vishwa Vidyapeetham*

FR.03.H3.2	Paper ID 222: Analysis of Resistive Switching Performance of ZnO/PVA Bilayer Memristive Device for Neuromorphic Applications	Saurabh Pandey *IIT Patna*
FR.03.H3.3	Paper ID 258: Study on Sensitivity Enhancement of SPR Optical Fiber Sensor via Hematite-Coated Au/Ag Grating	Anoop Pandey *IIT Bhubaneswar*
FR.03.H3.4	Paper ID 309: Single-Sided Optical Nanofiber Cavity: A Workbench for Cavity QED	Subrat Sahu *IIT Bhubaneswar*
FR.03.H3.5	Paper ID 391: Temperature Sensing Using Bare and Packaged Thermal FBGs: A Comparative Study	Rishika R *Cochin University of Science and Technology*
FR.03.H3.6	Paper ID 467: Defect Engineered in Carbon-Doped Hydroxyl-hBN Single Layer	Archana Singh *IIT Bhubaneshwar*
FR.03.H3.7	Paper ID 555: Comparison of strain measurements with optical fibre and conventional strain gauges	Sourav Sahoo *National Oceanography Centre, CSIR*
FR.03.H3.8	Paper ID 579: Multi-Periodic Structured Illumination Microscopy based on Co-linear Hexa Mirror geometry	Anupriya Tiwari *IIT Delhi*
FR.03.H3.9	Paper ID 593: High performance ultra-thin reduced graphene oxide photodetector for Visible light Detection at low Intensity and voltage range	Ravi Kumar Maddila *MNIT Jaipur*

Session – FR.04.H4.X

Advancements using AI
Chair: Dr. D. P. Dogra (IIT BBS) & Dr. Abhik Jana (IIT BBS)
Location: Hall 4 (L06S)
Time: 14:20 to 16:20

FR.04.H4.1	Paper ID 42: Music Genre Classification Using KNN Algorithm	Sasikumar Periyasamy *VIT*
FR.04.H4.2	Paper ID 46: Hyperparameter Optimization of CNN Models for Plant Leaf Disease Classification Using Optuna	Shantilata Palei *IIIT Bhubaneswar*
FR.04.H4.3	Paper ID 73: Unveiling Public Sentiment: A Twitter-Based Analysis of the 2024 Presidential Election	Tulika Chutia *Dibrugarh University*

FR.04.H4.4	Paper ID 121: Enhancing Road Safety with Real-Time Alerts	Mandi Sai Vinay *VR Siddhartha Engineering College*
FR.04.H4.5	Paper ID 186: Enhancing Book Discovery: Collaborative Filtering in Recommendation Systems	Shaina Sarika *Odisha University of Technology and Research*
FR.04.H4.6	Paper ID 266: GenAI Generated Audio Fraud Detection Using PCA and SVM	Jayanti Rout *Fakir Mohan University*
FR.04.H4.7	Paper ID 415: Bharat Park: A Comprehensive Analysis of Multi-factor Dynamic Pricing Strategies for Urban Parking and Congestion Reduction	Ishan Sharma *KIET Group of Institutions*

Session – FR.05.H5.X

Bioeffects and Medical Applications
Chair: Dr. Himanshu Padole (IIT BBS) & Dr. Nitya Tiwari (IIT BBS)
Location: Hall 5 (L07S)
Time: 14:20 to 16:00

FR.05.H5.1	Paper ID 41: Heart Disease Detection Using ECG Signal Based on Temporal Convolutional Networks Technique	Abhishek Kumar *CV Raman Global University*
FR.05.H5.2	Paper ID 107: Air Sense: Internet of Things-enabled Novel Power Efficient Indoor Air Quality Monitoring System	Radhika Raina *IIT ROPAR*
FR.05.H5.3	Paper ID 145: Efficient Brain Tumor Detection with a Lightweight Customized CNN Approach	Debendra Muduli *CV Raman Global University*
FR.05.H5.4	Paper ID 178: A Comparative Analysis of InceptionV3 and Xception Deep Learning Models for Enhanced Nail disease Diagnosis	Lalit Kumar Behera *Berhampur University*
FR.05.H5.5	Paper ID 366: An Ensemble Convolutional Neural Network -Bidirectional Long Short-Term Memory-Attention Approach for Electrocardiogram-Based Heart Disease Diagnosis	Soumyaranjan Patro *C.V. Raman Global University*
FR.05.H5.6	Paper ID 433: Gait Recognition Module Based on Metrics for Speed-Invariant Classroom Attendance System	Anusha R *Bapuji Institute of Engineering and Technology*

FR.05.H5.7	Paper ID 532: Millimeter Wave FMCW MIMO Radar for Remote Vital Sign Monitoring: Experimental Verification Using Empirical Wavelets and MUSIC Algorithm	Debalina Ghosh *IIT Bhubaneswar*

Conference Day 2: Saturday, 24th May 2025

Session – SA.06.H1.X

Novel Antenna Designs
Chair: Prof. A. R. Harish (IIT Kanpur)
Location: Hall 1 (L03S)
Time: 9:15 to 11:00

SA.06.H1.1	Paper ID 35: Metasurface Loaded Closely Spaced Shared Radiator 4-port MIMO Antenna with Dual-Radiation Performance	Jyotiprakash Swain *KIIT University*
SA.06.H1.2	Paper ID 320: Design of Low Profile Ultra Wideband Slotted Blade Antenna for Airborne Applications	Jitendra Kumar *CABS, DRDO*
SA.06.H1.3	Paper ID 339: Low Profile L-Band CP Omni-Directional Antenna for Line-of-Sight Service Link	Pramendra Verma *DEAL, DRDO*
SA.06.H1.4	Paper ID 344: Characterization and Evaluation of Stacked Patch Circularly Polarized Antenna for 5G n260 Applications	Munasa Yuvaraju *IIT Bhubaneswar*
SA.06.H1.5	Paper ID 405: Corrugated Directional SIW Linearly Tapered Slot Antenna with Novel Co-Axial Feeding for C- Band Application	Bharath Reddy Gudibandi *VIT*
SA.06.H1.6	Paper ID 427: Design and Development of a Triple-Band Microstrip Antenna Using Metamaterial and DGS for Sub-6 GHz Applications	Lakshman Kukkadapu *Vardhaman College of Engineering*
SA.06.H1.7	Paper ID 568: Multi-band wearable antenna for off-body applications for 5g nr bands with the bending analysis	Munasa Yuvaraju *IIT Bhubaneswar*

Session – SA.07.H2.X

AI in Communication System Design
Chair: Prof. Mihir N. Mohanty (ITER) & Dr. Soumya P Das (IIT BBS)
Location: Hall 2 (L04S)
Time: 9:15 to 11:00

SA.07.H2.1	Paper ID 213: Blockchain Based Multimedia Content Authentication Using Ethereum: Transformation Towards Decentralized Framework	Jayanti Rout *Fakir Mohan University*
SA.07.H2.2	Paper ID 214: Neuro-Symbolic AI: Pioneering Semantic Communication for 6G Wireless Networks	Faizan Shafi *IIT Dharwad*
SA.07.H2.3	Paper ID 288: Federated Learning Based Smart Radio Networks: Enabling Cross-Platform Communication	Jyoti Ranjan Satapathy *IIT Indore*
SA.07.H2.4	Paper ID 468: EEEAM-RSCA:A Novel Heuristic for Energy-Efficient Survivable Multipath Routing in SDM-EON	Sourabh Chandra *Netaji Subhash Engineering College*
SA.07.H2.5	Paper ID 541: A Novel AI-Powered Approach for Narrow Beam Management in 5G and Beyond Networks	Adarsh Ravi *IIT Patna*
SA.07.H2.6	Paper ID 557: Optimal 3D-Deployment of UAVs for Disaster Communication using Reinforcement Learning	Parakhee Brahma *IIT Bhubaneswar*
SA.07.H2.7	Paper ID 574: A Simplified Constraint Heuristic Optimization Algorithm for Energy Efficient Edge Server Placement in 5G Network	Satyasai Jagannath Nanda *MNIT Jaipur*

Session – SA.08.H3.X

Optimization for Advanced Systems
Chair: Dr. Sanjay Kumar Gowre (JSPM University Pune)
Location: Hall 3 (L05S)
Time: 9:15 to 11:00

SA.08.H3.1	Paper ID 294: Multi-User Multiple-Input Multiple-Output (MIMO) Communications Through Binary Flower Pollination Algorithm-Based User and Antenna Scheduling	Jyoti Mohanty *Siksha O Anusandhan University*

SA.08.H3.2	Paper ID 323: Lower order polynomial based trajectory generation for mobile robots	Snehil Sanyal *NIT Raipur*
SA.08.H3.3	Paper ID 341: Optical cryptosystem-based multimodal biometric security	Gaurav Verma *LNMIIT*
SA.08.H3.4	Paper ID 416: Characterization of an Encoder-only model on Heterogeneous Systems	Devashree Tripathy *IIT Bhubaneswar*
SA.08.H3.5	Paper ID 567: Optimizing Dual-Function Radar and Communication: A Performance Study of FMCW, OFDM, and OTFS	Tasmin Mishra *IIT Bhubaneswar*
SA.08.H3.6	Paper ID 587: MI-STAR: Mitigation of Self-Interference in Simultaneous Transmit and Receive Array via Evolutionary Optimization	Amisha Verma *IIT Bhubaneswar*
SA.08.H3.7	Paper ID 596: SFDR Improvement in DDS Using Memory Compression Techniques	JS Mangat *DLRL, DRDO*

Session – SA.09.H4.X

Advancements in Healthcare using AI
Chair: Dr. Veerendra D. (Guru Nanak Dev Engineering College) & Satyasai J. Nanda (MNIT Jaipur)
Location: Hall 4 (L06S)
Time: 9:15 to 10:30

SA.09.H4.1	Paper ID 9: Verification of Himalayan Meditative State Using EEG Rhythms	Pankaj Sahu *Dr B R Ambedkar NIT Jalandhar*
SA.09.H4.2	Paper ID 63: A Two-Stage Anomaly Detection Framework for Improved Healthcare Using Support Vector Machines and Regression Models	Chittaranjan Hota *BITS-Pilani, Hyderabad*
SA.09.H4.3	Paper ID 313: Deep Neural Network with Dilated Convolutions and MRI Imaging for Multi-Class Staging Prediction of Alzheimer's Disease Progression	Rakesh C Joshi *Amity University*
SA.09.H4.4	Paper ID 316: A Hybrid Approach for Emotions Classification using Modified Intrinsic Time Scale Decomposition	Priyadarsini Samal *NIT Warangal*
SA.09.H4.5	Paper ID 385: Improved Brain Tumor Diagnosis through Feature Selection and Hybrid Ensemble Models	Debendra Muduli *CV Raman Global University*

Session – SA.10.H1.X

Microwave and mm-Wave Sensors in Advanced Applications
Chair: Dr. Atri Mukhopadhyay (IIT BBS) & Dr. Varun Kumar (Shimane University, Japan)
Location: Hall 1 (L03S)
Time: 14:00 to 15:30

SA.10.H1.1	Paper ID 365: 3D LiDAR-Based Online Monitoring of Sodium Combustion Aerosols Concentration via Regression Analysis	Avinash Acharya *IGCAR, Homi Bhabha National Institute*
SA.10.H1.2	Paper ID 369: Novel Technique for UWB based Breathing Detection	Sandeep Acharya *Infineon Semiconductor Technologies*
SA.10.H1.3	Paper ID 389: Microwave Resonator Based Blood Glucose Sensor	Abhishek Kumar, NIT Warangal
SA.10.H1.4	Paper ID 501: Wi-Fi CSI based Human Activity Monitoring using Machine Learning and Deep Learning algorithms	Ranjit Kumar *IIT Bhubaneswar*
SA.10.H1.5	Paper ID 530: Bandwidth Enhancement of Slotted Hexagonal Patch Antenna for 6G Ultra-Fast Data Transfer and Brain Computer Interface	Uppala Rao *Bapatla Engineering College*
SA.10.H1.6	Paper ID 553: Microwave Sensing for Bone Density Evaluation: A Time-Frequency Perspective	Nishtha Shrivastava *IIT Bhubaneswar*

Session – SA.11.H2.X

Communication System Design and Optimization
Chair: Prof. P. Chakrabarti (IIEST Shibpur)
Location: Hall 2 (L04S)
Time: 14:00 to 15:45

SA.11.H2.1	Paper ID 204: A Novel Framework for Adaptive Jamming with Dynamic Null Zones for Securing Tactical Communications	Jyoti Ranjan Satapathy *Indian Army*
SA.11.H2.2	Paper ID 267: High Code Rate 16-QAM Golay Complementary Set Construction for Communication Systems	Aditya Prakash *IIT Patna*
SA.11.H2.3	Paper ID 305: UAV-Assisted IRS-SIMO System in 6G Communication	Nikita Goel *Manipal Institute of Technology*

SA.11.H2.4	Paper ID 384: Delay Domain Reuse for Multi-cell OTFS	Saif Khan Mohammed\ *IIT Delhi*
SA.11.H2.5	Paper ID 418: Design and Implementation of Molecular Communication (MC) Test Bed	Praneeth Mangala *VNIT Nagpur*
SA.11.H2.6	Paper ID 430: PIC based Implementation of DPS-QKD Transceiver	Ashay Atkar *VNIT Nagpur*
SA.11.H2.7	Paper ID 451: Enhancing Security through Correlation and Integration of Multi-Source Network Logs	Shivaram GV *Alva's Institute of Engineering and Technology*

Session – SA.12.H3.X

Sensing and Tracking Applications
Chair: Dr. Rakesh Balabantaray (IIIT Bhubaneswar)
Location: Hall 3 (L05S)
Time: 14:00 to 15:30

SA.12.H3.1	Paper ID 187: Adaptive Histogram Equalization with Fine-Tuned DNN-YOLOv9 Technique for Vehicle Detection in Adverse Weather	Ram C. Barik *C. V. Raman Global University*
SA.12.H3.2	Paper ID 299: Fiber Modal Interferometric Networks for Enhanced Vehicle Identification and Traffic Flow Analysis	Rajan Jha *IIT Bhubaneswar*
SA.12.H3.3	Paper ID 305: Laser-Based Detection of Structural Vibrations of a Model Truss Bridge	Ajit Jha *University of Agder, Norway*
SA.12.H3.4	Paper ID 325: Geo-tagging and Tracking for Detected Vehicles using UAV	Avani Vipat *Sardar Patel Institute of Technology*
SA.12.H3.5	Paper ID 402: Robust Vehicle Detection in Aerial Imagery: Comparative Study of YOLO Variants	Diya Kulkarni *Sardar Patel Institute of Technology*
SA.12.H3.6	Paper ID 576: Unitary-ESPRIT for 2D-DOD and 2D-DOA Estimation with EMV Sensors in MIMO Radar	Srinivasarao Chintagunta *NIT Calicut*

Session – SA.13.H4.X0

Communication and Sensing on IOT Platforms
Chair: Dr. Siddhartha Borkotoky (IIT BBS)
Location: Hall 4 (L06S)
Time: 14:00 to 15:45

SA.13.H4.1	Paper ID 44: Integrating IOT devices for Monitoring and Enhancing Working Conditions of Workers in India	Sasikumar Periyasamy, VIT
SA.13.H4.2	Paper ID 406: U-Net: A Binary Segmentation Architecture for Building Localization in Structural Health Monitoring using the Xbd Dataset	Sampa Sahoo, NIT Rourkela
SA.13.H4.3	Paper ID 409: Internet of Things to Steer Point-of-Care Testing of Vital Sign Monitoring	Richa Khandelwal *Shri Ramdeobaba College of Engineering and Management*
SA.13.H4.4	Paper ID 506: Enhanced GIFT Encryption with Chaotic Maps for Secure Data Transmission in LoRa-Based IoT Devices	Balakrishnan Ramalingam *VIT Chennai*
SA.13.H4.5	Paper ID 569: A Novel Approach for an Enhanced Security System on IoT Devices	Karthik P *Cambridge Institute of Technology*
SA.13.H4.6	Paper ID 577: Design and Development of a novel multi-functional ESP32-based smart device featuring Optical Communication, Photometric Analysis, and Real-Time Data Processing	Ural Oinam *IIT Madras*
SA.13.H4.7	Paper ID 595: React.js and Django REST Framework Architecture for IoT Applications	Shankar S S *C-DAC, Thiruvananthapuram*

Conference Day 3: Sunday, 25th May 2025

Session – SU.14.H1.X

Metasurfaces for Field Shaping and Synthesis
Chair: Prof. S. Mahapatra (IIT Kharagpur) & Dr. Srinivasarao C. (NIT Calicut)
Location: Hall 1 (L03S)
Time: 10:15 to 12:30

SU.14.H1.1	Paper ID 345: A Concave Circular Brick-Shaped Metamaterial-Inspired Superstrate Based High-Gain Microstrip Patch Antenna for Wideband Applications	Uday Kumar *BIT Mesra*

SU.14.H1.2	Paper ID 388: Design of Chalcogenide Phase change metamaterial-based microwave absorber	Atchyut M *IIT Madras*
SU.14.H1.3	Paper ID 400: Conformal Metasurface based Linear to Circular Polarization Converter for Dual Band Wi-Fi Applications	Sai Sreenija Yallanki *IIT Kanpur*
SU.14.H1.4	Paper ID 408: A Frequency-Agile Filtering Metasurface Antenna with Single, Dual, and Wideband Reconfigurability	Sai Sharanya Vattam *NIT Warangal*
SU.14.H1.5	Paper ID 542: Capacitive-loaded Ultra-thin Polarization insensitive Dual-band Metamaterial Absorber	Punyatoya Routray *IIT Bhubaneswar*
SU.14.H1.6	Paper ID 543: Reconfigurable Rasorber with Wide Absorption Band and Switchable Refection/Transmission Band for 5G and Bluetooth Applications	Punyatoya Routray *IIT Bhubaneswar*
SU.14.H1.7	Paper ID 566: Design and Optimization of a 5G Conformal Microwave Absorber for the n78 Band	Priyanshu Mathur *IIT Bhubaneswar*
SU.14.H1.8	Paper ID 580: Polarization-Insensitive Reflective Metalens at the telecommunication Wavelength: Design and numerical study	Jagriti Ahuja *IIT Delhi*
SU.14.H1.9	Paper ID 588: A 1-Bit Reconfigurable Reflective Metasurface for Sub-6 GHz Applications	Vaibhav Srivastava *IIT Bhubaneswar*

Session – SU.15.H2.X

Data Processing and AI for Imaging Applications
Chair: Prof. A. K. Ojha (IIT BBS) & Dr. E. Sivasankar (NIT Trichy)
Location: Hall 2 (L04S)
Time: 10:15 to 12:15

SU.15.H2.1	Paper ID 274: A Zero-Reference based HE + Deep Enhancement Approach for Low-Light Images	Abhipsa Mahala *C. V. Raman Global University*
SU.15.H2.2	Paper ID 301: Metaheuristics-based Parameter Optimization of Bilateral Gamma Correction for Image Enhancement Applications	Simi R *Manipal Institute of Technology*
SU.15.H2.3	Paper ID 311: DeepDementia: A Deep Neural Network Integrated Convolutional Block Attention Module for Multi-Stage Dementia Classification from Brain Imaging	Rakesh C. Joshi *Amity University*

SU.15.H2.4	Paper ID 373: Advancements in Image Captioning: A Comparative Analysis of Embeddings, Attention Mechanisms, and Transformer-Based Models	Venkatesh A *NIT Trichy*
SU.15.H2.5	Paper ID 399: Enhancing Intracranial Hemorrhage Classification Using Feature Fusion in Brain Stroke Imaging	Mudassir Rafi *SRMAP*
SU.15.H2.6	Paper ID 436: Exploring Geospatial Data: Fundamentals and Challenges in Change Detection	Minal Apsangi *Nirma University*
SU.15.H2.7	Paper ID 514: Image Compression using Neural Networks	Abhinav Sharma *IIT Guwahati*
SU.15.H2.8	Paper ID 549: Offline Signature Verification Using CNN and Siamese Networks: A Comparative Analysis	Snehasis Nayak *KIIT University*

Session – SU.16.H3.X

Advances in Optical Communication ***Chair:*** Dr. Bijayananda Patnaik (NIT Raipur) & Dr. Ravi K. Maddila (MNIT Jaipur) ***Location:*** Hall 3 (L05S) ***Time:*** 10:15 to 12:15		
SU.16.H3.1	Paper ID 488: Implementation of SDON for Industrial Internet and Industry 4.0 in Google Colab	Sadia Sara *Meghnad Saha Institute of Technology*
SU.16.H3.2	Paper ID 538: Study on Photonics-Based Microwave Signal Generation Scheme for 5G Application	Shreepreet Sahu *IIT Bhubaneswar*
SU.16.H3.3	Paper ID 550: Performance Studies of Inter-Satellite Optical Wireless Communication Links Using PPM	Nalluri Satwika *IIT Bhubaneswar*
SU.16.H3.4	Paper ID 551: Machine Learning-Based Classification of OAM Modes in FSO Systems Under Fog Conditions Using SVM and KNN	Bijayananda Patnaik *NIT Raipur*
SU.16.H3.5	Paper ID 556: Design and study of an OAM-based advanced multiplexing technique for FSO communication system under turbulent channel conditions	Subhrajyoti Sunani *IIT Bhubaneswar*
SU.16.H3.6	Paper ID 558: Study on Pointing Error and Atmospheric Turbulence in FSO for 5G and Beyond Communication	Prasant K. Sahu, IIT Bhubaneswar

SU.16.H3.7	Paper ID 559: Free-Space Optical Communication: Navigating the Potential of Photon Entanglement and 16-QAM	Prasant K. Sahu, IIT Bhubaneswar
SU.16.H3.8	Paper ID 565: Reliability of an FPGA-Based Air-to-Water-to-Air VLC Link: Experimental Investigations	Vedant Borde, IIT Kharagpur

Session – SU.17.H4.X

Sensing for Human Machine Interface ***Chair:*** Prof. B. Sundarajan (IISC Bangalore) ***Location:*** Hall 4 (L06S) ***Time:*** 10:15 to 11:45		
SU.17.H4.1	Paper ID 128: Kannada Speech to Sign Language Converter: A Novel Platform Integrates Indian Sign Language and Speech-to-Text Technology	Pushpa Latha *Ramaiah Institute of Technology*
SU.17.H4.2	Paper ID 149: Model Agnostic Few Shot Learning for Finger Vein Recognition	Kiranjit Pattnaik *KIIT University*
SU.17.H4.3	Paper ID 182: Object Detection for BVI people in Low Light Conditions using improved TBEFN Technique	Shraddha Gugulothu *YCCE*
SU.17.H4.4	Paper ID 217: Efficacy Prediction of Low Observable Measures with Range & Search Time Using Eye Tracker	Amit Bhati *DLJ, DRDO*
SU.17.H4.5	Paper ID 307: Contrastive Loss Coupled with Occlusion Aware Training Aids in Face Recognition from Low Quality Group Photos	Sam Narayana Subudhi *IIT Bhubaneswar*
SU.17.H4.6	Paper ID 485: Person Tracking System	Keerthana P *Reva university*

Session – SU.18.H5.X

Biomedical Imaging and Sensors ***Chair:*** Dr. N. B. Puhan (IIT BBS) & Dr. R P Padhy (IIT BBS) ***Location:*** Hall 5 (L07S) ***Time:*** 10:15 to 11:45		
SU.18.H5.1	Paper ID 218: CMOS Capacitor-Coupled Based Instrumentation Amplifier For ECG Signal Acquisition	Prabhat Barik, IIT Bhubaneswar

SU.18.H5.2	Paper ID 310: Fabry-Perot based Flexible System for Human Pulse Monitoring	Pratik Mishra *IIT Bhubaneswar*
SU.18.H5.3	Paper ID 364: Enhanced Multi-Class Brain Tumor Classification Model using MRI images: A Wavelet and Deep Learned Features based Approach	Debendra Muduli *CV Raman Global University*
SU.18.H5.4	Paper ID 411: Design a compact UWB antenna for the detection of breast tumor via Non-Invasive Microwave Imaging	Tarun Panwar *Thapar Institute of Engineering and Technology*
SU.18.H5.5	Paper ID 419: WatershedSE-UNet: A Hybrid Deep Learning and Morphological Approach for Lung Nodule Segmentation in CT Scans	Jiddu Krishnan O P *NIT Silchar*
SU.18.H5.6	Paper ID 592: Deep Spatio-Temporal Networks for Visual Speech Recognition: A Hybrid CTC-Cross Entropy Approach	Nikhil Niranjan *IIT Bhubaneswar*

ABSTRACT OF CONTRIBUTORY SESSIONS

Contributory Sessions 1

Session FR.01.H1.X

ELECTROMAGNETIC MODELLING OF DEVICES AND SYSTEMS

48

Design and Realization of Gysel Power Divider & Combiner for High Power Handling in C-Band

Manish Mishra

Bharat Electronics Limited
*E-mail: manishmishra@bel.co.in**

This paper describes the design and realization of a Gysel power divider & combiner capable of pulse RF power handling up to hundreds of watts in C-band. The designed & realized micro strip Gysel power divider (GPD) has an insertion loss better than 3.6dB over the frequency band 5.4-5.9GHz. The measured bandwidth for 15dB return loss is 8.8% while the isolation between two output ports is better than 19dB. Good agreement between simulated and measured results is seen. Further two GPDs have been used to generate more than 500W of pulsed power in the C-band by combining 2 solid state pulsed power amplifiers with good stability over the frequency band. Roger 6035 HTC with design dielectric constant of 3.6 with 20mils thickness is used to implement GPD.

* Corresponding author

259

Design and Simulation of C-Band 500 MW Relativistic Backward Wave Oscillator

Sanjay Gupta

MTRDC, DRDO

E-mail: sanjay.mtrdc@gov.in

Relativistic backward-wave oscillator (R-BWO) in C-Band frequency designed and simulated using the "CST 3D electromagnetic software-BWO is high efficiency and high power microwave sources[1]based on backward-wave oscillations and very high energy relativistic electron beams are able to generate high-power microwave radiation. Particle-in-cell simulation is carried out to study for the beam-wave interaction in a Relativistic Backward Wave Oscillator, using the 3D finite integration-based particle-in-cell (PIC) code (CST). The model includes a corrugated slow wave structure along with cavity reflector and collector, and electron beam contained by solenoid focusing. All the components of the model were simulated in three dimensions in order to study the effects of overall electromagnetic fields on Corrugated SWS circuit and electron beam interaction. In this paper we explain the design and PIC Simulation of C-band 500MW Over mode corrugated coaxial R-BWO in which a coaxial corrugated waveguides used. The frequency of R-BWO can be tune by varying beam voltage, this structure having very high-power generating capability and is capable to produce higher peak power.

262

Large signal GaN HEMT Model up to Ku Band for Switch Applications

Kirti Bansal

SSPL, DRDO

E-mail: kirti86@gmail.com

In this paper, the nonlinear model of RF switch based on measurement data of indigenous fabricated devices up to Ku Band is presented. The GaN switch devices are fabricated with 0.25 μm Gallium Nitride (GaN) High Electron Mobility Transistor (HEMT) technology on a 100 μm SiC substrate. The model includes the determination of various parameters like insertion loss, return loss, isolation, maximum output power, saturated power, and 1 dB compression point. These parameters are extracted and validated with the measurement results. Good agreement between the modelled and measured parameters has been achieved up to Ku band. Using this large signal model, maximum power handling has been estimated which is approximately 39 dBm for 0.6 mm device.

329

Precision Phase Matched Low Noise Figure High Gain C Band RF Down Converter

Narendra S

Bharat Electronics Limited Bangalore
E-mail: narendras@bel.co.in

This paper describes the design and development of a compact ultra-low Noise Figure, high gain, precision phase matched C band RF down converter used in the receive chain of a Monopulse Radar which converts a C-Band RF signal to an IF signal. The module has integrated gain and phase matching provision to match between independent receivers. The module has a gain of greater than 50dB with a gain matching specification of ±1 dB and phase matching of ±10° across frequency.

332

Graphene-Inspired THz Frequency-Reconfigurable Antennas for Biomedical Applications

Sangeetha Subbaraj[1], Shweta Thomas[2*], Aniket Kumar Singh[2], Robin Philip[2] & Tanmay Mishra[2]

[1]*College of Engineering, Guindy, Anna University, Chennai*
[2]*Vellore Institute of Technology, Vellore, Tamil Nadu, India*
E-mail: sangeetha6077@gmail.com; shwetabthomas04@gmail.com; aniketkumar.singh2022@vitstudent.ac.in; robin.philip2022@vitstudent.ac.in; tanmay.mishra2022@vitstudebt.ac.in

The increasing demand for high-resolution biomedical imaging, non-invasive diagnostics, and advanced wireless healthcare applications has driven significant interest in terahertz (THz) technology. Graphene, with its exceptional electrical and mechanical properties, has emerged as a promising material for designing frequency-reconfigurable antennas in the THz band.

This paper presents a novel graphene-inspired THz frequency-reconfigurable antenna tailored for biomedical applications. The proposed antenna leverages graphene's tuneable conductivity to achieve dynamic frequency reconfiguration, enabling real-time adaptability to different biomedical scenarios such as targeted sensing, spectroscopy, and wireless body-area networks (WBANs). A detailed design methodology is provided, incorporating electromagnetic simulations to evaluate key performance metrics, including gain, bandwidth, and tunability. The results demonstrate that the graphene-based reconfigurable antenna offers enhanced efficiency, miniaturization, and improved impedance matching compared to conventional metallic antennas. Furthermore, we explore the potential integration of this technology into wearable and implantable biomedical devices. The findings highlight the feasibility of graphene-enabled THz antennas as a transformative solution for next-generation biomedical sensing and communication systems.

350

Examining the Possibility of rGO@PANI with co-doped Mn_2O_3 and $CoFe_2O_4$ Composites as a Radar-absorbing Material for the whole X and Ku Bands

Parveen Kumar* **& Ashavani Kumar**

National Institute of Technology Kurukshetra
E-mail: parveen16021998@gmail.com, ashavani@yahoo.com

Reduced graphene oxide (rGO) with diverse, layered structures and unique electronic characteristics was restricted in microwave wave absorption owing to the high dielectric and easy agglomeration properties. A series of rGO@ PANI-decorated co-doped with Mn_2O_3 and $CoFe_2O_4$ was synthesized via the in-situ polymerization method. By altering the reactant ratio, the composite's ferrite content and impedance matching can be more precisely controlled. The subsequent rGO@PANI co-doped with Mn_2O_3 and $CoFe_2O_4$ (PGF) hybrids combine the dielectric loss and magnetic loss characteristics while retaining the layered structure of the rGO@PANI nanosheets. The resultant material is thoroughly characterized for phase and morphological analysis, dielectric, and magnetic parameters, which revealed that Mn_2O_3 with $CoFe_2O_4$ particles were dispersed across the rGO@PANI composite. Our finding shows that the 35 wt. % loading sample PGF 1 shows an effective absorption bandwidth (EAB) of 5.07 GHz with RLmax –22.79 dB at 2 mm thickness; for PGF 2, the achieved RLmax is -48.44 dB at 2.3 mm thickness with 3.6 GHz of EAB; while PGF 3 exhibits an EAB of 2.64 GHz and RLmax of –47.03 dB at 7.28

GHz with a thickness of 3.4 mm. The notable differences in the conduction of electrons and electronic structure between the decorated Mn_2O_3 and $CoFe_2O_4$ particles and the outer graphene layers significantly enhance the interfacial polarization. By creating quicker electron transport routes, the 3D hierarchical rGO@PANI with co-doped Mn_2O_3 and $CoFe_2O_4$ network topology improves absorption outcomes. The well-thought-out planning and the cooperative action of rGO@PANI sheets with co-doped Mn_2O_3 and $CoFe_2O_4$ are intimately connected to high interfacial polarisation, which results in incredibly efficient absorption with lightweight and thin thickness as suitable in the field of RADAR absorption and for the protection of space gadgets from electromagnetic pollution. Remarkably, PGF 3 with 40 wt. % loading shows the absorption of microwaves for the entire X and Ku bands only at a thickness of 2.4 mm and covers 92.5% of the whole frequency region (2-18 GHz) up to a thickness of 6 mm and can be utilised for outstanding performance in defence applications and in 5G/sub-6G communications.

495

High Efficiency High Output and Wideband GaN HEMT Power Amplifier over Entire S-Band

Pradeep Gorre[1*], Sampath Kumar Bandi[1], Aswini Kumar Samantaray[2] & Sandeep Kumar Amera[3]

[1]*Vignan's Foundation for Science, Technology & Research*
[2]*Manipal Institute of Technology Bengaluru*
[3]*Special Centre for Nanosciences JNU*
E-mail: pradeepgorre@gmail.com; sampathkumar1219722@gmail.com; aswini.samantaray@manipal.edu; fedrer.engg@gmail.com

This work proposes a high Output Power, high efficiency and wideband GaN HEMT based power amplifier (PA) in S-band frequency for high efficiency and wideband applications. The design methodology employs a U-shaped Matching network (UMN) at input and output sections to suppress harmonics impedances which improves the i/o power, efficiency, and bandwidth over S-band with minimal power consumption The work employs, a Class AB Power Amplifier integrated with the proposed UMN is designed using 0.15 μm GaN HEMT. The simulated results proves that the proposed design achieves a high power added efficiency (PAE) of 49%–68%, output power, Pout, of 39 dBm–44.5 dBm, drain efficiency (DE) of 54%–80%, and a nearly flat gain small-signal gain of 13.8 dB over a frequency of 2 GHz-4GHz.

512

Standing Wave Pattern and Frequency Measurement System using Capacitively Coupled Probes for Pedagogical Applications

Debapratim Ghosh*, Nitish Pani & Mahendra Kheti

Indian Institute of Technology Bhubaneswar
E-mail: debapratim@iitbbs.ac.in; a23ec09011@iitbbs.ac.in; 23sp06006@iitbbs.ac.in

In this article, a microstrip standing wave pattern and frequency measurement system is proposed and developed for pedagogy relating to a first undergraduate engineering course on electromagnetic waves. The proposed system helps in capturing the positional power along a transmission line which helps in visualizing the standing wave. Thereafter, from these measurements, the frequency of the traveling wave is also calculated with a reasonably good accuracy of 3–5%. The proposed system is suitable for UHF frequencies and is tested over the frequency range from 400 to 600 MHz the proposed system is an alternative to network analysers and slotted lines which help students visualize the standing wave phenomena and also perform frequency calculations.

564

Design of a Wideband Millimeter-Wave 5th Order Chebyshev CPW Bandpass Filter

Manshree Mishra, Ashish Kumar & Ashwini Sawant*

IIT Guwahati
E-mail: manshree.mishra@iisi.iitg.ac.in; ashish2003@iitg.ac.in; ashwinisawant@iitg.ac.in

This paper describes a wideband coplanar-waveguide (CPW) bandpass filter (BPF) employing open-circuited stubs for millimeter-wave (mmW) applications. A conventional design approach of fifth-order 0.1 dB Chebyshev BPF is utilized to implement it using open-circuited stubs. Spurious modes suppression is carried out by patterning a low resistance path on either side of the stubs over a two-layered substrate. The design achieves 30% bandwidth with a low insertion loss of 1.69 dB.

Contributory Sessions 2

Session FR.02.H2.X

PROPAGATION AND CHANNEL MODELLING

165

Channel Estimation for Intelligent Reflecting Surfaces for MISO-OFDM System by Deep Learning

SUBHRA SAHU* & POONAM SINGH

National Institute of Technology, Rourkela

E-mail: subhrajyotisahu123@gmail.com; psingh@nitrkl.ac.in

The Intelligent Reflecting Surface (IRS) framework improves next-generation wireless communication systems by using affordable passive components. Because Long Short-Term Memory (LSTM) models can learn temporal correlations and extract features from sequential data, they are very effective at solving Channel Estimation (CE) problems. This work applies the LSTM model to CE in downlink Orthogonal Frequency Division Multiplexing (OFDM) systems with IRS-assisted multiple-input single-output (MISO). The Deep Neural Network (DNN) optimization technique is used to train the LSTM model, and comparison is done using available DNN optimizers. The CE performance is assessed using various optimizers under different cyclic prefix lengths. For IRS assisted communication systems, simulation findings validate the LSTM model's capacity to learn efficiently and offer accurate channel estimation.

183

Ensemble Learning-based Holistic Attack Classification for Enhanced Physical Layer Security

SWETHAMBRI M

Anna University

E-mail: swethambrisundaram@gmail.com

The wireless medium is vulnerable to various security threats. Ensuring confidentiality, availability and authentication is crucial for secure communication. Physical Layer Security (PLS) utilizes the physical properties of the communication medium. It enhances wireless security by incorporating ML algorithms to accurately detect and classify different attacks. This paper proposes a novel approach to classify various attacks such as spoofing, eavesdropping and Denial of Service (DoS) using a holistic Hybrid Learning (HL) approach. The proposed HL based ensemble method consists of combination of probabilistic and Boosting-based classifiers, which improves the overall attack classification accuracy. The generated dataset contains various attack and non-attack scenarios. The proposed technique extracts significant features from the data and these features are fed to the ensemble model, which effectively distinguishes between different attack and non-attack scenarios. The simulated results show that the proposed model achieves an improved classification accuracy of 91%, with precision, recall, and F1-score of all achieving greater than 90%. The proposed HL technique with high accuracy reliably distinguishes between different types of attacks for the generated datasets.

208

Modulation Classification: Diverse Techniques and Constellation Model Training at Varying Resolution

Varun Kumar*, Md. Ganim & Sakshi Somani

IIIT Vadodara

E-mail: varun_kumar@diu.iiitvadodara.ac.in; 202211048@diu.iiitvadodara.ac.in; 202491002@diu.iiitvadodara.ac.in

In both wired and wireless scenarios, modulation techniques play a crucial role in enhancing QoS, ensuring robustness against fading channels, reducing BER, and supporting various applications. In this work, various models, including CNN, LSTM, and ResNet, were trained on the RadioML dataset with IQ time-series, amplitude-phase, and constellation diagrams as input features. Our results indicate that time-series features with the hybrid CNN-LSTM model outperformed traditional CNN models, particularly under low SNR conditions, while amplitude-phase time-series enhanced classification performance at high SNRs. However, performance across datasets with varying channel conditions showed limited adaptability. Notably, combining time-series and constellation inputs demonstrated potential for enhanced generalization, paving the way for robust real-world modulation recognition.

215

Analytical ABEP and Diversity Order of IRS-Assisted Wireless Systems in Nakagami-m Fading

Sanjeet Bhagat

IIT Bhubaneswar
E-mail: skb18@iitbbs.ac.in

In this paper, we employ an intelligent reflecting surface (IRS) to enhance the performance of the communication system using GRQAM signalling. We derive a precise average bit error probability (ABEP) formulation for coherent, Gray-coded rectangular quadrature amplitude modulation and non-coherent modulation techniques, valid for a larger number of reflecting elements of the IRS. We derive both the generalized and the asymptotic ABEP expressions employing a moment-generating function-based technique. The diversity order is obtained from ABEP equations, and it depends on the number of passive reflecting elements on the IRS and fading parameter. Our results show that ABEP performance decreases when the fading parameter or the number of reflecting elements increases. Monte Carlo simulation is used to verify the accuracy of our obtained analytical ABEP equations.

251

A Novel Precoder for Peak-to-Average Power Ratio Reduction in OTFS Systems

Saurabh Prakash[1], Venkatesh Khammammetti[2] & Saif Khan Mohammed[3*]

[1]*Department of Electrical Engineering, Indian Institute of Technology Delhi)*
[2]*Department of Electrical and Computer Engineering, Duke University*
[3]*Department of Electrical Engineering, Indian Institute of Technology Delhi*
E-mail: Saurabh.Prakash@ee.iitd.ac.in; venkatesh.khammammetti@gmail.com; saifkmohammed@gmail.com

We consider the issue of high peak-to-average-power ratio (PAPR) of Orthogonal time frequency space (OTFS) modulated signals. This paper proposes a low-complexity novel iterative PAPR reduction method which achieves a PAPR reduction of roughly 5 dB when compared to a OTFS modulated signal without any PAPR reduction. Simulations reveal that the PAPR achieved by the proposed method is significantly better than that achieved by other state-of-art methods. Simulations also reveal that the error rate performance of OTFS based systems with the proposed PAPR reduction is similar to that achieved with the other state-of-art PAPR reduction methods.

276

A Lightweight Video Streaming Framework for Low-End Systems Using FFMPEG and UDP for Real-Time Transmission

Neha Gupta

Bharat Electronics Limited
E-mail: neha201004@gmail.com

Live streaming allows us to watch, create, and share videos in real time, similar to live TV, making seamless video transmission essential. In applications like face recognition systems, the system requires an input stream, processes it, and then generates an output stream. Current live video streaming methods often use the OpenCV library, which consumes a high percentage of CPU resources. This paper introduces an algorithm for live video streaming using the FFMPEG library and UDP socket. The algorithm operates in two phases: the first phase functions as a stream generator, while the second phase acts as a stream receiver. It is applicable to both live CCTV video feeds and pre-recorded data. Our approach using the FFMPEG streamer library significantly reduces CPU and memory usage.

348

A Comparative Analysis of Advanced Modulation and Coding Schemes for LEO Based Satellite Network

Deepika Manjunath

Bharat Electronics Limited
E-mail: deepikamrevankar@gmail.com

Satellite communication using LEO is gaining popular acceptance, owing to its capabilities of ubiquitous coverage, high speed, and minimum latency. The aforementioned advantages offered by LEO links are best utilized when the adverse challenges such as Doppler shifts and varying channel conditions observed in LEO communication are effectively addressed. In order to ensure reliable connectivity amidst these challenges, it is pertinent to employ optimum Modulation and Coding Schemes (MCS). A comparative analysis of the various advanced MCS provides necessary insights into the trade-offs between spectral efficiency, BER, throughput and the effectiveness with which a reliable communication link can be established under dynamic LEO channel conditions.

353

Higher Order Time-Frequency Uncertainty Products and Self-Dual Equations

VINEET GHULE[1*], SUKETU PATNI[1], SHAILESHRAM SIVAKUMAR[1], VIKRAM GADRE[1], MURALI PAVULURI[1], MANISH SHARMA[2], SHRIKANT SHARMA[1&3] & RADHAKRISHNA P.[3]

[1]*IIT Bombay*
[2]*IITRAM Ahmedabad*
[3]*LRDE Bengaluru*
E-mail: vineetghule17@gmail.com; 23b1299@iitb.ac.in; 23b1205@iitb.ac.in; vmgadre@ee.iitb.ac.in; pavuluri.murali1@gmail.com; manishsharma@iitram.ac.in; shrikantsharma75@gmail.com; prk.lrde@gmail.com

The time-frequency uncertainty principle is a key concept in signal processing that delineates the trade-off between a function's distribution in time and frequency domains. This principle is vital across various fields, including communications, quantum mechanics, and control theory, where signals that are optimally localized in both time and frequency are of paramount importance. In this chapter, we build upon the classical uncertainty principle by investigating time-bandwidth products related to higher-order moments, thereby broadening the understanding of time-frequency localization. We analyse these higher-order uncertainty products for particular families of functions, such as Gaussians and splines, and delve into their minimization. The exploration of optimal functions leads us to self-dual ordinary differential equations (ODEs), which are commonly encountered in time-bandwidth optimization scenarios. We characterize these self-dual ODEs, elucidate their properties through the two-sided Laplace transform, and devise strategies for their resolution. By linking time-frequency uncertainty, function optimization, and self-duality, this chapter offers valuable insights into essential signal analysis challenges and their wider significance in engineering and applied mathematics.

Contributory Sessions 3

Session FR.03.H3.X

OPTICAL SENSORS AND DETECTORS

88

Assessing Fiber Bragg Grating Sensors for Landslide Monitoring: Current Trends and Future Prospects

Ajish U

Amrita School of Engineering and Technology
E-mail: imaj94@rediffmail.com

The effectiveness of Fiber Bragg Grating (FBG) sensors in Landslide monitoring is analysed in this review article, which also highlights some possible future avenues for these sensors. Advanced monitoring methods are required for early detection and risk mitigation since landslides pose serious dangers to human safety and infrastructure. FBG sensors have become important instruments in geotechnical monitoring because of their great sensitivity and capacity to sense displacement, temperature, and strain. The fundamentals of FBG technology are covered in the article, the paper also discusses the difficulties in deploying FBG sensors in various geological settings and highlights the need for more study to improve their performance and incorporation into early warning systems. This work attempts to give insights into the current practices of FBG sensor applications in landslide monitoring by synthesizing existing literature and case studies. It also outlines future research directions to increase the efficacy of FBG sensors in hazard prediction and management.

222

Analysis of Resistive Switching Performance of ZnO/PVA Bilayer Memristive Device for Neuromorphic Applications

SAURABH PANDEY

IIT Patna

E-mail: saurabh@iitp.ac.in

The development and analysis of a resistive switching device with an Ag/PVA/ZnO/ITO architecture for memory applications are examined in this article. Utilizing the complementary advantages of metal oxides and organic polymers, the device incorporates a bilayer of zinc oxide (ZnO) and polyvinyl alcohol (PVA) as the active switching medium to improve device performance. The PVA/ZnO bilayer simulates synaptic activity, with the Ag electrode Configured as the upper electrode while the ITO electrode is the underlying electrode. Bipolar analog resistive switching (ARS), which is typified by slow resistance changes that mimic potentiation and depression in biological synapses, is confirmed by current-voltage (I-V) analysis. Ion migration and the creation and breaking of conductive filaments inside the bilayer are the causes of these characteristics. The results advance our knowledge of the device's switching mechanisms and synaptic emulation capabilities while highlighting its potential for low-power, scalable neuromorphic systems.

258

Study on Sensitivity Enhancement of SPR Optical Fiber Sensor via Hematite-Coated Au/Ag Grating

ANOOP PANDEY*, DEBALINA GHOSH & PRASANT KUMAR SAHU

Indian Institute of Technology Bhubaneswar

E-mail: ap39@iitbbs.ac.in; deghosh@iitbbs.ac.in; pks@iitbbs.ac.in

This study explores the optimization of a D-shaped Surface Plasmon Resonance (SPR) sensor for accurate measurement of external refractive index (RI). The sensitivity is analysed by varying the RI from 1.33 to 1.39, the residual cladding thickness (dr) from 0.1 μm to 1 μm, the Ag layer thickness from 40 nm to 65 nm, and the Fe_2O_3 layer thickness from 3 nm to 14 nm. Based on the response analysis, the optimal parameter values are identified. The proposed D-shaped SPR optical fiber sensor achieves an exceptional sensitivity of 15,000 nm/RIU, surpassing previously reported values.

309

Single-Sided Optical Nanofiber Cavity: A Workbench for Cavity QED

Subrat Sahu[1*], Kali P. Nayak[2] & Rajan Jha[1]

[1]*IIT Bhubaneswar*

[2]*Department of Engineering Science, University of Electro-Communications, Chofu-Shi, Tokyo 182-8585, Japan*

E-mail: ss99@iitbbs.ac.in; kalipnayak@uec.ac.jp; rjha@iitbbs.ac.in*

We propose a single-sided optical nanofiber cavity (SSONFC) to facilitate cavity quantum electrodynamics (cQED). The structure is designed and analysed through three-dimensional finite-difference time-domain simulations. A systematic evaluation of the cavity properties and spontaneous emission characteristics has been conducted. Our findings demonstrate that this design significantly enhances the spontaneous emission of a single quantum emitter (SQE). A Purcell factor of ~ 22 and a maximum unidirectional single photon coupling efficiency of around 72 % into the ONF's guided mode can be achieved when an SQE is positioned on the SSONFC surface. Additionally, a high scattering limited quality factor (Q ~ 2510) is obtained for the fundamental cavity mode. The structure's dimensions can be adjusted to accommodate different SQE emission wavelengths, making it a versatile solution. This fiber-integrated single-photon device holds potential for applications in quantum information processing, quantum phase switches, and single-photon transistors.

391

Temperature Sensing Using Bare and Packaged Thermal FBGs: A Comparative Study

Rishika R

Cochin University of Science and Technology

*E-mail: rishikar@ug.cusat.ac.in**

This experiment investigates the temperature sensitivity of Fiber Bragg Gratings (FBGs) by analysing their wavelength shift in response to variation in temperature. Two types of FBGs are used, a bare FBG and a packaged thermal FBG. The results demonstrate a consistent and repeatable pattern, indicating that the packaged thermal FBG exhibited a higher sensitivity to temperature changes compared to the bare FBG. Graphical analysis confirmed the linear relationship between temperature change and wavelength shift. This study demonstrates the effectiveness of FBGs for precise temperature sensing, highlighting their potential in industrial and scientific applications.

467

Defect Engineered in Carbon-Doped Hydroxyl-hBN Single Layer

Archana Singh

IIT Bhubaneshwar

E-mail: pdfph24002@iitbbs.ac.in

Hexagonal boron nitride (h-BN) is a layered material that is garnering interest in a variety of fields due to its intriguing characteristics. We describe a density functional theory approach to carbon defects, which introduces a novel method for accurately engineering the electrical characteristics of hydroxyl-functionalized hBN monolayers using strategic carbon doping. We use a novel method to investigating impurity states in our C-doped h-BN by building it into van der Waals heterostructures with other 2D materials. By substituting carbon atoms for nitrogen atoms, we enable the production of self-assembled hexagonal carbon rings at the doping location, lowering the bandgap from 5.53 eV to 2.02 eV. This change effectively converts the material from an insulator to a semiconductor, which has great promise for practical applications.

555

Comparison of Strain Measurements with Optical Fibre and Conventional Strain Gauges

Sourav Sahoo

National Oceanography Centre

*E-mail: sahoosoton@gmail.com**

Distributed optical fibre sensing (DOFS) transforms a cable into an array of sensors, which can be used to detect and monitor multiple physical parameters such as temperature, vibration and strain, with high spatial and temporal resolution over long distances (up to 100s of km). DOFS has the potential to transform our ability to gather critical data about our environment and infrastructure, through use of optical fibres within legacy and/or new cables, installed for bespoke science applications. DOFS offers low cost, high sensitivity, real-time, continuous monitoring with crucially non-intrusive and passive environmental monitoring attributes, which significantly reduces the carbon footprint compared to conventional point sensors, like seismometers, CTD (Conductivity, Temperature, & Depth sensor), ADCPs, that involve significant ship costs for deployment and recovery. Conventional sensors for temperature, strain and elastic waves provide point measurements and typically do not relay that information in real-time. Juxtapose DOFS, it can transform a cable into a distributed dense array of many hundreds or thousands of sensors that all provide information instantaneously. Over the past few years, DOFS has gained immense interest in geophysical monitoring, with several high-impact publications demonstrating its potentially wide-reaching application to hazards assessment and environmental monitoring1,2,3. DOFS offers certain benefits over conventional seismic sensors such as ease and cost of deployment on the seafloor or downhole with restricted access, for a much higher number of equivalent point sensors. Although there is a need of calibration between DOFS and seismometer signals to enable independent measurements with DOFS. Here, using state of the art rock physics laboratory facilities, we present results of calibration on typical geological materials, from reservoir rocks to marine sediments, to accurately compare sensor measurements so we can relate them with certainty to future studies of seafloor seismo-acoustic wave propagation phenomena and their applications. We have just started this project and here we have included some initial results.

579

Multi-Periodic Structured Illumination Microscopy Based on Co-linear Hexa Mirror Geometry

Anupriya Tiwari

Indian Institute of Technology, Delhi
E-mail: anupriyatiwari671@gmail.com

Structured illumination microscopy (SIM) is a robust widefield optical nanoscopic technique. Several approaches are implemented to improve SIM's resolution capability (sim-fold). However, achieving high-resolution with large field-of-view (FOV) is still challenging. We present a tilt-mirror based multi-periodic SIM for large-FOV super-resolution microscopy. The sample is illuminated by multi-periodic structured pattern generated by six-beam interference using a custom-designed mirror-mount. We achieve 3.16-fold resolution improvement while using 20X/0.40 numerical-aperture objective that supports large-FOV (0.53 times 0.34mm^2). This overcomes high-space bandwidth product challenge, achieving 9.98-fold improvement. The mMP-SIM decouples illumination and collection paths, enabling scalable super-resolution over large-FOV. By using 28X/0.80 objective lens, optical resolution of 170 nm over (0.40 times 0.25) mm^2 imaging area is demonstrated. The proof-of-principle experimental demonstration is performed both for fluorescent beads as well as bio-sample like U2OS (human bone osteosarcoma) cells.

593

High Performance Ultra-thin Reduced Graphene Oxide Photodetector for Visible Light Detection at Low Intensity and Voltage Range

Ravi Kumar Maddila

MNIT Jaipur

E-mail: rkmaddila.ece@mnit.ac.in

A reduced graphene oxide-based photodetector has been fabricated for visible light detection by using drop casting process. the synthesis of reduction graphene oxide was accomplished chemically by hydrazine hydrate. The surface and I-V characteristics were analysed by Raman spectroscopy, TGA, FESEM and XRD. A thin film of 1 μL rGO photodetector was fabricated on silicon-based substrate with conducting fingers of gold. Further, the effect of variation of incident power (1-80 μW) and applied biased voltage (1-5 volts) on current generated have been analysed at 550 nm. the response time and responsivity has been measured for wide power and voltage range. It was found possible to achieve high responsivity (0.12 A/W) with less time response at very low radiation intensity (1 μW) and low voltage.

Contributory Sessions 4

Session FR.04.H4.X

OPTICAL SENSORS AND DETECTORS

42

Music Genre Classification Using KNN Algorithm

Kapila Shobit, Aayush Vishnoi, Jayishnu Agarwal, Kritin Bhardwaj, K S Kerti, Anupurba Guha, Sasikumar Periyasamy* & Maheswari S

VIT, Chennai

E-mail: kapila.krishnabhaskar2021@vitstudent.ac.in; aayush.vishnoi2021@vitstudent.ac.in; jayishnu.agarwal2021@vitstudent.ac.in; kritin.bhardwaj2021@vitstudent.ac.in; kerti.ks2021@vitstudent.ac.in; anupurba.guha2021@vitstudent.ac.in; sasikumar.p@vit.ac.in; maheswari.s@vit.ac.in

An Acoustic Source Identification (ASI) approach for machine learning-based music genre recognition is presented in this paper. We train a K-Nearest Neighbour (KNN) model by extracting important acoustic features from a variety of audio datasets. Results from experiments show that this method works well, categorizing different types of music with great accuracy and precision. The performance of the method is influenced by KNN's simplicity, resistance to noise and interference, and dependability.

46

Hyperparameter Optimization of CNN Models for Plant Leaf Disease Classification Using Optuna

Shantilata Palei & Puspanjali Mohapatra

IIIT, BBSR/CSE

E-mail: c122011@iiit-bh.ac.in; puspanjali@iiit-bh.ac.in

Leaf diseases are a major challenge for agricultural productivity, requiring accurate and efficient detection methods. This study proposes a robust approach for multi-class classification of leaf diseases in apple and mango plants using MobileNetV2 and XceptionNet. A diverse dataset enhances model generalizability, while the Optuna framework optimizes hyperparameters for improved performance. Among the tested models, Optuna-optimized

MobileNetV2 achieves the highest accuracy of 94.05%. The comparative analysis highlights the advantages of combining advanced CNN architectures with hyperparameter optimization for leaf disease detection. This study emphasizes the role of optimized models in addressing complex agricultural challenges and their potential to support data-driven decision-making. The findings pave the way for future research deploying such techniques for scalable and real-time smart farming solutions.

73

Unveiling Public Sentiment: A Twitter-Based Analysis of the 2024 Presidential Election

Tulika Chutia

Dibrugarh University
E-mail: tulikachutiadu24@gmail.com

The US presidential election of 2024 has ignited debates among candidates and ideas on social media, especially on Twitter. The paper determines whether a comment is positive, negative, or neutral through the analysis of the language used in each comment based on deep learning techniques, namely convolutional neural networks (CNN) and recurrent neural networks (RNN). The study is conducted on public opinion in the 2024 US presidential election on Twitter data that are collected manually. Based on the accuracy, precision, recall, and F1-score values, the CNN model performed better than the RNN model. For the same measures, the RNN model produced 93%, 93.10%, 92.79%, and 92.87% for accuracy, precision, recall, and F1-score. The results demonstrate how well deep learning models classify sentiment and provide insight into how the general public views political events. Future work will focus on integrating multilingualism, transfer learning, and attention mechanisms along with Transformer models.

121

Enhancing Road Safety with Real-Time Alerts

Mandi Sai Vinay*, Sukanya P, Leela Sri Harshavardhan Vallabhani, Eshitha Kondapalli & Jaswanth Vadde

Velagapudi Ramakrishna Siddhartha Engineering College
E-mail: saivinaymandi828@gmail.com; sukanya@vrsiddhartha.ac.in; vlsharshav@gmail.com; eshithakondapalli@gmail.co; jaswanthvadde9@gmail.com

Road unsafety and crimes are the major issues now a days in India itself as we can observe every day many accidents, thefts and some serious crime incidents have happened. In this work, we have designed a mobile application that offers real-time alerts based on the status and location of travellers in relation to accident-prone, theft-prone, or crime-prone areas. The application taps into historical incident data from police stations and works with the Google Maps API to map locations deemed high risk, at 100m radius, alerts a user that is coming something of these highlighted areas, gets alerted for the road users and can take necessary precautions. To improve the situation, the app aims to enhance road users to be aware of anything that could lead to accidents or thefts and other crimes, thus helping its users have a safer travel experience. Mobile Application Road Safety with Real-Time Alerts using Google Maps API in India.

186

Enhancing Book Discovery: Collaborative Filtering in Recommendation Systems

Shaina Sarika*, Ashis Mishra & Neelima Mohanty

Odisha University of Technology and Research
E-mail: shainasarika@gmail.com; akmishracse@outr.ac.in; neelimagoogly@gmail.com

The rapid rise in customer demand has significantly contributed to the growth of information technology and e-commerce. This surge in demand is largely driven by the widespread use of web services that offer personalized user experiences, especially through recommendation systems. These systems utilize advanced information filtering techniques to suggest items based on user preferences, past behaviours, and shared similarities among users. Recommendation systems consider various factors such as price, quality, publisher, author, and other attributes to deliver highly relevant suggestions. In the context of books, such systems are invaluable for students, researchers, and avid readers, enabling them to discover new material aligned with their unique interests. Book recommendation systems leverage collaborative filtering, which focuses on identifying patterns in user behaviours and preferences to make more accurate recommendations. By analysing user profiles, search history, and purchasing trends, these systems can present a diverse array of books tailored to individual needs. For students and academics, this technology can greatly enhance the discovery of relevant literature, providing suggestions that not only match current preferences but also expose users to new topics and authors. The integration of collaborative filtering into book recommendation systems thus plays a crucial role in improving decision-making, broadening the scope of exploration, and fostering a more personalized reading experience. This research will delve into how collaborative filtering enhances book discovery and benefits various user demographics.

266

GenAI Generated Audio Fraud Detection Using PCA and SVM

Jayanti Rout1*, Minati Mishra1 & Ram Chandra Barik2

[1]*P. G. Department of Computer Science, Fakir Mohan University, Balasore, Odisha*
[2]*Departments of CSE, C.V. Raman Global University, Bhubaneswar, Odisha*
E-mail: routjayanti2k19@ieee.org; minatiminuu@gmail.com; ram.chandra@cgu-odisha.ac.in

As Artificial Intelligence (AI) continues to advance at a remarkable pace, the widespread production of synthetic audio has led to concerns about authenticity and security. This study presents an effective machine learning-based detection method using MFCC, Mel spectrograms, and an SVM classifier. The proposed approach extracts key spectral features from audio signals, standardizes them, and applies Principal Component Analysis (PCA) to reduce dimensionality while preserving 95% of the variance. A Support Vector Machine (SVM) with a linear kernel is used for classification, ensuring robustness and accuracy. To guarantee a fair and reliable evaluation, a 5-fold stratified cross-validation strategy is implemented. The results indicate that the model effectively differentiates real and AI-generated audio with high accuracy, surpassing traditional methods. This research advances digital content authentication and helps counter the risks posed by AI-generated audio manipulation.

415

Bharat Park: A Comprehensive Analysis of Multi-factor Dynamic Pricing Strategies for Urban Parking and Congestion Reduction

Ishan Sharma*, Harsh Rajput, Arunima Shukla, Astha Gupta & Meeta Chaudhry

KIET Group of Institutions
E-mail: ishan.2125csit1105@kiet.edu; harsh.2125csit1058@kiet.edu; arunima.2125csit1097@kiet.edu; astha.2125csit1083@kiet.edu; chaudhrymeeta1@gmail.com

BharatPark is a real-time solution to parking issues and underutilization of resources. Unwanted congestion from parking deficiencies leads to on-street problems. A user-friendly app allocates parking spaces via vendors. It is a two-user authenticated application, one for vendors and the other for users (Parkers). Built on Flutter+Firebase, it includes dynamic pricing to manage congestion, guided map locations, residential parking, and more. The back-end stores user/vendor information, order details, and timings, with dynamic pricing based on time, congestion, and vehicle type (two-wheeler or four-wheeler). A key feature addresses on-street parking by allowing off- street vendors to blacklist vehicles, storing their details in the back end.

Contributory Sessions 5

Session FR.05.H5.X

BIOEFFECTS AND MEDICAL APPLICATIONS

41

Heart Disease Detection Using ECG Signal Based on Temporal Convolutional Networks Technique

Abhishek Kumar*, Runku Dilleswara Rao, Adyasha Rath, Ganapati Panda & Sujit Chandra Das

CV Raman Global University Odisha

E-mail: abhishek8520kr@gmail.com; dilleswararaorunku123@gmail.com; adyasha.rath@cgu-odisha.ac.in; ganapati.panda@gmail.com; sujitchandra2077@gmail.com

As a major cause of death worldwide, Heart Disease (HD) underscores the critical demand for reliable and accurate diagnostic techniques. Traditional methods of analysing electrocardiogram (ECG) signals rely. heavily on manual interpretation, which is prone to human error and often delays diagnosis. This study presents an automated approach for HD detection using Temporal Convolutional Networks (TCNs), Utilizing their capability to model and understand long-term relationships within sequential data. The MIT-BIH Arrhythmia dataset was utilized, with preprocessing steps including segmentation and oversampling to address class imbalance between normal (N) and abnormal (A) heartbeats. The TCN model, incorporating causal convolutions, dilation, and residual connections, demonstrated robust performance in distinguishing between N and An ECG signal. The system displayed a 96% test accuracy (Acc), alongside impressive precision (Prec) and recall (Re) scores, particularly in the A class. This research emphasizes how TCNs can improve both the Acc and efficiency of HD detection, making them a reliable and scalable option for early diagnosis. This work contributes to the ongoing effort to integrate machine learning (ML) into healthcare, aiming to reduce the burden of cardiovascular diseases through timely intervention.

107

Air Sense: Internet of Things-enabled Novel Power Efficient Indoor Air Quality Monitoring System

Radhika Raina

IIT Ropar

*E-mail: radhika.20eez0022@iitrpr.ac.in**

Air pollution, known to cause health issues such as respiratory and cardiovascular diseases emphasize the importance of monitoring indoor air quality with carbon dioxide (CO_2) sensor. Despite various air quality monitoring systems being discussed in the literature, their power consumption is often overlooked. This paper introduces Internet of Things (IoT) enabled novel power efficient indoor air quality monitoring system that uses CO_2, temperature and humidity sensors which send data to the cloud via Global System for Mobile Communication (GSM) technology. Moreover, keeping CO_2 levels below 1000 PPM is vital for optimal indoor air quality as discussed in literature. The CO_2 sensor operates continuously, while the temperature and humidity sensor activate when the CO_2 level becomes equal or exceeds 1000 PPM. By combining temperature and humidity data with CO_2 levels, a more thorough insight into air quality is achieved. This strategy will reduce the power consumption of the device as both the sensors are active when there is a critical need of monitoring. The system's battery life is approximately 9.71 days when CO_2 sensor is active only, decreasing to 8.71 days when both sensors are active using a 15600 mAh / 4.2 V battery. Furthermore, the sensor's fetching interval is set to 15 min when CO_2 values are unstable. However, when CO_2 values remain stable, the fetching interval starts at 15 min and doubles with each successive stage of continued stability. This approach minimizes the device's power consumption by reducing the frequency of data collection.

145

Efficient Brain Tumour Detection with a Lightweight Customized CNN Approach

Shantanu Shookdeb, Purnima Pattnaik, Rahul Maurya, Debendra Muduli* & Santosh Kumar Sharma

CV Raman Global University

E-mail: shookdebshantanu@gmail.com; pattnaikpurnima903@gmail.com; raahulmaurya2@gmail.com; muduli.debendra@gmail.com; santoshsilu84@gmail.com

Brain tumour detection is a critical task in medical imaging, requiring accurate and efficient classification methods. In this work, we propose a lightweight customized Convolutional Neural Network (CNN) model designed to achieve high accuracy with minimal computational complexity. Our model significantly reduces the number of tuneable parameters while maintaining robust feature extraction capabilities. Experimental results demonstrate that the proposed approach achieves an impressive classification accuracy of 98.03%, outperforming several predefined deep learning models. The reduced computational overhead makes the model suitable for real-time and resource-constrained environments, such as edge devices and mobile healthcare applications. This study highlights the potential of customized CNN architectures in advancing brain tumour diagnosis, ensuring both efficiency and accuracy in medical image analysis.

178

A Comparative Analysis of InceptionV3 and Xception Deep Learning Models for Enhanced Nail Disease Diagnosis

Lalit Kumar Behera

Berhampur University

E-mail: lkb.rs.cs@buodisha.edu.in

Nail diseases, often early indicators of systemic health conditions, require accurate and timely detection for effective management. Traditional diagnostic methods are often time-consuming and require expert intervention, making automated detection systems increasingly desirable. Deep learning techniques, particularly convolutional neural networks (CNNs), have shown promise in the field of medical image analysis. This study presents a comparative analysis of two advanced deep learning models, InceptionV3 and Xception, for the detection and classification of nail diseases. Both models, known for their performance in image recognition tasks, were evaluated based on their accuracy, efficiency, and robustness in detecting various nail diseases from a dataset of nail images. The results demonstrate that while both models offer competitive performance, Xception outperforms InceptionV3 in terms of accuracy and model stability, particularly in distinguishing between subtle differences in disease categories. This study highlights the potential of deep learning models in enhancing diagnostic tools for dermatology and contributes valuable insights into selecting optimal architectures for medical image classification tasks. Future research may explore model optimization and expansion to other skin and nail-related conditions for more comprehensive automated diagnostic systems.

366

An Ensemble Convolutional Neural Network -Bidirectional Long Short-Term Memory-Attention Approach for Electrocardiogram-Based Heart Disease Diagnosis

Soumyaranjan Patro*, Rahul Sharma, Gourav Palai, Siddhi Mishra, Adyasha Rath & Ganapati Panda

C. V. Raman Global University

E-mail: patrosoumya2003@gmail.com; rahulsharma2rk2329@gmail.com; palaigourav99@gmail.com; siddhiprakashmishra08@gmail.com; adyasha.rath@cgu-odisha.ac.in; ganapati.panda@gmail.com

The global prevalence of cardiovascular diseases demands urgent and precise electrocardiogram (ECG) analysis for proper diagnosis because these diseases lead to numerous deaths each year. The proposed research presents an ensemble deep learning framework which utilizes convolutional neural networks (CNNs) with bidirectional long short-term memory (BiLSTM) layers combined with attention mechanisms to identify temporal and spatial patterns in raw ECG recordings. The preprocessing step includes normalization followed by overlapping segmentation of multi-channel ECG signals to extract dynamic signal patterns. Individual ensemble members use the combination of two-stage CNN architecture with residual components for local feature extraction followed by BiLSTM component that models temporal dependencies and an attention layer that enhances discriminative sequence parts before fully connected layers produce the final classification. The ensemble approach demonstrates strong performance on the PTB Diagnostic ECG Database by reaching high accuracy coupled with precision and recall and F1 scores as well as superior results than single-model architectures. Through deep feature extraction with CNNs with sequential modelling using BiLSTM along with an attention mechanism for providing salient feature weights the approach provides better generalization and reduced overfitting. The framework unites various modelling methods to establish a possible real-time ECG-based diagnostic system which can improve reliable cardiovascular disease screening in medical environments.

433

Gait Recognition Module Based on Metrics for Speed-Invariant Classroom Attendance System

Anusha R[*1] & Shridhar Sanshi[2]

[1]*Bapuji Institute of Engineering and Technology, Karnataka*
[2]*National Institute of Technology Karnataka, Surathkal*
E-mail: anur.research@gmail.com; shridhar.sanshi@gmail.com

Gait, how a person walks, is a critical biometric feature utilized in various human detection applications, including systems for tracking classroom attendance. The system's ability to detect gait in these scenarios is influenced by speed, as student entry speed can differ depending on various factors. This research presents a method for minimizing the effects of speed changes on a gait detection system. The initial step involves extracting two features, named regional similar metric and regional dissimilar metric, using similar and dissimilar regions from the training and test samples. After that, a metric-based classification is performed to categorize a test sample with a training sample. This method enhances gait detection performance by effectively recognizing slight changes in static data. The CASIA C and OU-ISIR A gait datasets are employed in assessing this approach. Compared to other methods, this gait recognition technique shows promising results based on the experiment conducted.

532

Millimeter Wave FMCW MIMO Radar for Remote Vital Sign Monitoring: Experimental Verification using Empirical Wavelets and MUSIC Algorithm

Debalina Ghosh*, Tasmin Mishra & Prasant Sahu

IIT Bhubaneswar

E-mail: deghosh@iitbbs.ac.in; a24ec09011@iitbbs.ac.in; pks@iitbbs.ac.in

Human vital sign monitoring is an important aspect of human health and wellness. In recent times, radar systems have found increased application in the domain because of its non-contact nature. In this study, a millimeter wave radar is used for human target range measurement and heart rate and breathing rate estimation. For this purpose, a multilevel combination algorithm is proposed and verified using experimental data. The study also compares the usefulness of the time-frequency estimation method involving Empirical wavelets vis-à-vis the eigenspace analysis method of multiple signal classification. The MIMO property of the radar is utilized for increasing the accuracy of the estimation method.

Contributory Sessions 6

Session SA.06.H1.X

NOVEL ANTENNA DESIGNS

35

Metasurface Loaded Closely Spaced Shared Radiator 4-port MIMO Antenna with Dual-Radiation Performance

Jyotiprakash Swain*, Asutosh Mohanty, Subhrakanta Behera & Wriddhi Bhowmik

KIIT University

E-mail: jyoti157ctc@gmail.com; asutoshmohanty.kiit0409@gmail.com; skbeherafet@kiit.ac.in; wriddhi.bhowmikfet@kiit.ac.in

A slot-based configured closely spaced shared radiator was utilized to serve as 4-port multiple-input-multiple-output(MIMO) antenna. Two distinct slots cavity were used to excite the radiating patch. The center-fed elliptical closed slot cavity provides magnetic field coupling while the open-ended diagonal slots provide electric field coupling. The induced coupling effect effectively improves the boresight radiation gain by top loading a metasurface layer. With the aforementioned design configuration, the proposed antenna asserts dual-radiation characteristics for Phi = 0deg and Phi = 90deg radiation planes. With the shared footprint, the antenna achieves better port isolation of > 20 dB, and the 10-dB impedance matching was improved by arranging the unit pixels of metasurface spacing. The prototype fabricated antenna of dimension 60mm x 60mm x 3.1mm was validated to operate from (4.31–5.85) GHz with 30.3% 10-dB IBW with maximum realized gain of 8.28 dBi. The performance merits show its viability for Super-high frequency (SHF) applications.

320

Design of Low Profile Ultra-Wideband Slotted Blade Antenna for Airborne Applications

Jitendra Kumar

CABS DRDO

E-mail: jitendra_k@ece.iitr.ac.in

This work presents a blade monopole antenna design and modelling operating in the 250–6000 MHz range. Because of their lightweight design and aerodynamic shape, blade antennas are the recommended antenna configuration for airborne applications. This design provides ultra-wide band bandwidth by slotting the blade on the flat surface. The dimensions of the antenna element are 376 × 352.9 mm. Antenna radiation pattern, gain, radiation efficiency, and resemblance to VSWR all accord well. The new slotted blade antenna design's low-profile structure and performance are commendable. The suggested frequency range has VSWR< 2 according to the antenna matching performance. The whole band exhibits gain > 0, with the simulated antenna gain at 6000 MHz being 8.5 dBi.

339

Low Profile L-Band CP Omni-Directional Antenna for Line-of-Sight Service Link

Pramendra Verma[1*], Raj Kumar[1], V K Singh[1], Preeti Verma[2] & V S Pandey[2]

[1]*DEAL (DRDO)*

[2]*NIT Delhi*

E-mail: pramendra.deal@gov.in; raj.kumar.deal@gov.in; vinod.singh.deal@gov.in; preetiverma@nitdelhi.ac.in; vspandey@nitdelhi.ac.in

A compact, low profile, lightweight, wide beam circularly polarized (CP) omni-directional antenna has been designed and developed for L-Band Service link for airborne platforms. A combination of a magnetic dipole and an electric monopole is used to generate a circular polarization. To achieve a compact form factor antenna, a PCB-based printed Alford loop with a top-loaded monopole configuration has been utilized. The antenna consists of a circular printed radiating element on RT Duroid 5880 dielectric substrate, four cylindrical shorting pins, metallic ground plane, Rohacell spacer, and Teflon based radome. The antenna was simulated and optimized using

ANSYS's HFSS electromagnetic simulation tool, with simulated results closely matching measured values. VSWR (<2) and axial ratio (<3dB) have been achieved over the frequency range of 980-1040 MHz and 940-1600 MHz, respectively. The antenna has a stable radiation pattern with 3600 azimuth and 800 elevation beam coverage. Due to its low-profile structure, it is highly suitable for airborne applications.

344

Characterization and Evaluation of Stacked Patch Circularly Polarized Antenna for 5G n260 Applications

Munasa Yuvaraju[1*], Anoop Pandey[1], Priyanshu Mathur[1], Aggala Naga Jyothi[2] & Debalina Ghosh[1]

[1]*IIT Bhubaneswar*

[2]*Vignan's*

E-mail: rajyr47@gmail.com; ap39@iitbbs.ac.in; 21ec01044@iitbbs.ac.in; aggala.jyothi@gmail.com; deghosh@iitbbs.ac.in

This manuscript details the write of a circularly polarized (CP) antenna which is suitable in design and favourable for the 5G performance domain. The anticipated antenna adopts a Rogers RT5880 substrate, designed into the dimensions of 5 mm × 5 mm × 0.5 mm. By introducing a rectangular cross-slot parasitic patch placed on a foam spacer over the standard patch antenna configuration, a remarkable gain of 8.32 dBi over the isotropic at 36.4 GHz is achieved. This novel technique is promising in terms of performance and flexibility for future 5G applications which constantly require wireless communication systems to be tailored to new parameters and challenges.

405

Corrugated Directional SIW Linearly Tapered Slot Antenna with Novel Co-Axial Feeding for C- Band Application

Bharath Reddy Gudibandi

VIT AP University
E-mail: bharathreddy.gudibandi@vitap.ac.in

Traditionally SIW H-plane horn antennas have less impedance matching when the thickness of the substrate is less than λ/10. Most of the existing H-plane horn designs are having operating frequency greater than 10GHz, because as operating frequency increases thickness of the substrate will not be a big problem. we designed a simple corrugated SIW linearly tapered slot antenna fed by novel coaxial feeding with FR-4 Epoxy as substrate; of thickness 0.0245λ and without any impedance matching structures. The proposed design gives combined benefits of horn and slot operating from 4.6 to 8GHz having peak measured gain up to 6.636dB.

427

Design and Development of a Triple-Band Microstrip Antenna Using Metamaterial and DGS for Sub-6 GHz Applications

Lakshman Kukkadapu*, Sulakshana Chilukuri, Shaik Jani Pasha & Tharun Madasu

Vardhaman College of Engineering
E-mail: lakshmankukkadapu@gmail.com; sulakshana@vardhaman.org; janipashajp522@gmail.com; tharunmadasu2004@gmail.com

This work presents the design and analysis of a triple-band microstrip antenna utilizing an Artificial Magnetic Conductor (AMC) metamaterial unit Cell. The antenna operates at frequencies of 3.27 GHz, 3.87 GHz, and 4.7 GHz with return losses exceeding 15 dB, achieving gains of 7 dB, 6.5 dB, and 10.4 dB at these frequencies, respectively. The antenna demonstrates bandwidths of 330 MHz, 460 MHz, and 240 MHz for each band, alongside a Front to Back Ratio (FBR) greater than 20 dB. The structure consists of two flag-shaped patches mounted on an FR4 substrate, with a Defective Ground Structure (DGS) incorporated to enable triple-band functionality. To further enhance the antenna's performance, an optimized AMC unit cell is introduced, improving gain and ensuring multiband operation. This design offers a high-performance solution with superior return loss, gain, bandwidth, and FBR in the Sub-6 GHz frequency range.

568

Multi-band Wearable Antenna for Off-body Applications for 5g nr Bands with the Bending Analysis

Munasa Yuvaraja*, Debalina Ghosh & Nishtha Shrivastava

IIT Bhubaneswar

E-mail: rajyr47@gmail.com; deghosh@iitbbs.ac.in; a23ec09006@iitbbs.ac.in

This is a multi-band wearable antenna tailored for 5G New Radio (NR) bands, specifically n65, n256, n79, and n47. The design is based on a straightforward microstrip patch and functions in various modes, mainly T001, T002 and T003, allowing it to resonate at three different frequencies. With compact dimensions of 58 × 56 × 2.2 mm^3 (0.46λ × 0.44λ × 0.017λ at 2.4 GHz), the proposed structure guarantees stable performance across all targeted bands. The antenna consistently achieving a gain of over 4 dB at all the resonating frequency bands, showcasing efficient radiation characteristics. Furthermore, a thorough bending analysis at angles of 10°, 15°, and 20° confirms its mechanical strength and minimal performance loss when deformed. Suitable for both on-body and off-body wearable applications, this antenna provides a dependable and efficient solution for next-generation wireless communication, ensuring optimal performance in practical wearable scenarios.

Contributory Sessions 7

Session SA.07.H2.X

AI IN COMMUNICATION SYSTEM DESIGN

213

Blockchain Based Multimedia Content Authentication Using Ethereum: Transformation Towards Decentralized Framework

Jayanti Rout[1*], Minati Mishra[1], Swatisipra Das[1] & Ram Chandra Barik[2]

[1]*P. G. Department of Computer Science, Fakir Mohan University, Balasore, Odisha*
[2]*Department of CSE, C. V. Raman Global University, Bhubaneswar, Odisha)*
E-mail: routjayanti2k19@ieee.org; minatimiinu@gmail.com; ssdas2022@gmail.com; ram.chandra@cgu-odisha.ac.in*

With the rapid growth of digital multimedia content and sophisticated adversary technologies, ensuring authenticity and security has become a critical challenge. Traditional methods, such as digital watermarks and cryptographic signatures, help verify authenticity. However, they come with various limitations, including challenges in key management, reliance on central authorities, and the implicit trust placed in these authorities, which can undermine security and authenticity. To address these challenges, this article presents a blockchain-based approach using Ethereum for the authentication of multimedia content. Using Ethereum's secure and transparent ledger, we create an immutable record of ownership and modifications, ensuring that digital content remains tamperproof and verifiable. This approach eliminates the need for centralized control, reduces the complexity of key management, and enhances security. Our analysis shows that the use of Ethereum smart contracts improves the reliability of digital content authentication, making it more secure and decentralized. Experiment reveal that our executed smart contract consumes 0.0011 SepoliaETH to store the metadata in blockchain and for authentication verification it consumes zero gas. This research highlights how the Ethereum blockchain can provide a robust and transparent solution to protect multimedia content, ensuring its integrity and authenticity in a trust less environment.

214

Neuro-Symbolic AI: Pioneering Semantic Communication for 6G Wireless Networks

Faizan Shafi[1*], Ashwin Waghmare[1], Rahul Pandya[1] & Sridhar Iyer[2]

[1]*Indian Institute of Technology Dharwad*
[2]*KLE Technological University, Dr. MSSCET*
E-mail: ee23dp004@iitdh.ac.in; 210010060@iitdh.ac.in; rpandya@iitdh.ac.in; sridhariyer1983@klescet.ac.in

The Sixth Generation (6G) wireless networks aim to enhance data transmission efficiency and reliability by adopting semantic communication, which prioritizes contextually relevant information over precise bit-level accuracy. This approach reduces bandwidth usage while preserving the intended meaning. This article presents a neuro-symbolic Artificial Intelligence (AI) based framework that integrates Logic Tensor Networks (LTNs) and Matrix Factorization (MF) to optimize the semantic encoding and decoding processes. By leveraging logical reasoning and shared knowledge, the proposed model efficiently reconstructs incomplete or distorted data, even in noisy transmission environments. Experimental evaluations, conducted across a range of Signal-to-Noise Ratios (SNRs), demonstrate the system's robustness in maintaining semantic reliability. Specifically, for low SNR values, the system demonstrated a decline in performance, with F1 scores between 0.52 and 0.57. However, as the SNR increased, the system's performance improved significantly, with the F1 score reaching up to 0.65, and recall values exceeding 0.78 for higher SNR levels. High specificity (0.96–0.97) across all the SNR values further emphasizes the system's ability to filter out irrelevant content effectively. These findings indicate that the proposed model is adaptable, scalable, and capable of efficiently transmitting context-rich information, making it suitable for a wide range of next-generation applications.

288

Federated Learning Based Smart Radio Networks: Enabling Cross-Platform Communication

Jyoti Ranjan Satapathy

IIT, Indore

E-mail: jyoti.ranjan1365@gmail.com

Modern communication demands intelligent radio grids for both military and civilian use. Specially during operations involving multiple countries or multiple agencies, legacy and current radio systems lack compatibility due to parameter differences. This leads to exchange of sets between teams to ensure communication, which is wastage of critical communication resources. Further, upgrading the legacy radio sets for Mobile Ad-Hoc Networks (MANET) and cognitive radio systems requires costly hardware replacements. This paper proposes a framework to enhance wireless communication efficiency using both contemporary and legacy radios. It employs LightGBM based hierarchical model in a Federated Learning (FL) architecture for RF classification. Integrated with Software Defined Radio (SDR), it dynamically adjusts radio parameters for real-time interoperability and MANET formation. The solution enables different radio makes and models to work together, creating a smart radio grid. Spectrum sensing, signal classification, automatic parameter adaptation, spoofing attack mitigation, and fingerprint-based authentication are the key features of this solution. Experiments show a 92% accuracy in radio classification and full interoperability across different bands and models. This improves efficiency, reduces costs, and enhances system performance.

468

EEEAM-RSCA:A Novel Heuristic for Energy-Efficient Survivable Multipath Routing in SDM-EON

Sourabh Chandra

Netaji Subhash Engineering College

E-mail: chandra.sourabh@gmail.com

The exponential growth of data traffic in optical networks necessitates scalable and efficient solutions. Space Division Multiplexing Elastic Optical Networks (SDM-EON) offer a promising approach by utilizing multi-core fibers

(MCFs) to expand capacity. However, SDM-EONs introduce challenges such as inter-core crosstalk, spectral fragmentation, and high energy consumption. Traditional routing, spectrum, and core assignment (RSCA) strategies address these issues independently, leading to suboptimal performance. This paper presents Enhanced Energy-Efficient Adaptive Multipath RSCA (EEEAM-RSCA), a heuristic that integrates energy-aware multipath routing, crosstalk-minimized core selection, and fragmentation-resilient spectrum allocation. Extensive simulations on NSFNET, COST239, and USA Backbone topologies demonstrate that EEEAM-RSCA achieves a 20% reduction in bandwidth blocking ratio (BBR), 18% lower energy consumption, and 22% improved fragmentation efficiency compared to state-of-the-art approaches such as CMDE-RSCA and FMDE-RSCA. These results highlight EEEAM-RSCA as a robust and practical solution for future SDM-EON deployments.

541

A Novel AI-Powered Approach for Narrow Beam Management in 5G and Beyond Networks

Divyansh Gupta, Aruj Gautam, Adarsh Ravi* & Preetam Kumar

Indian Institute of Technology Patna

E-mail: divyansh_2101ee89@iitp.ac.in; aruj_2101ee15@iitp.ac.in; adarsh_2021ee01@iitp.ac.in; pkumar@iitp.ac.in

Future 5G and beyond networks will demand fast, energy-efficient, and secure narrow beam selection to maintain robust connectivity after the initial access phase. In this research, we leverage artificial intelligence (AI), and machine learning (ML) based methods for secure narrow beam selection that predicts the optimal narrow beam between the base station and user equipment (UE), maximizing connectivity while minimizing computational complexity. Our AI/ML models, evaluated through confusion matrices and top-k accuracy metrics, demonstrate outstanding performance, with a top 1 accuracy exceeding 90% for k-nearest neighbours (KNN), random forest (RF), and deep neural networks (DNN). A key novelty of this work is the incorporation of scatterer locations along with receiver locations as input features to the AI/ML model. While previous studies have discussed the importance of scatterer information, its implementation has not been explored extensively. By leveraging this innovative feature set, our model enhances prediction accuracy and provides a robust solution for beam management in 5G advanced networks.

557

Optimal 3D-Deployment of UAVs for Disaster Communication using Reinforcement Learning

Parakhee Brahma

IIT Bhubaneswar
E-mail: 23sp06001@iitbbs.ac.in

Unmanned Aerial Vehicle (UAV)-assisted wireless communication networks have garnered substantial research attention in recent years, particularly as solutions for establishing communication in areas affected by natural disasters. The rapid deployment capabilities and mobility of UAVs make them ideal for enhancing the Quality of Service (QoS) in such scenarios. In a disaster-stricken region, users spread changes unpredictably in real time, causing changes in network demand. Consequently, UAV deployment strategies must continuously adapt to provide better emergency services. The geographical area under consideration is divided into optimal, non-overlapping clusters, each served by a cluster UAV, with a relay UAV linking it to the central base station (CBS). The QoS considered here is to maximize the user coverage with limited available resources. To achieve the QoS, a solution based on reinforcement learning is proposed where the problem is solved in three parts. First, determining the cluster formation and UAV placement. Next, optimize the cluster UAV height and power. Finally, determining the optimal relay UAV trajectory with minimal travel distance. Computer simulations demonstrate the effectiveness of the proposed approach, outperforming random deployment and convex optimization methods in achieving better QoS.

574

A Simplified Constraint Heuristic Optimization Algorithm for Energy Efficient Edge Server Placement in 5G Network

Loveena Singh, Satyasai Jagannath Nanda* & Rajendra Prasad Yadav

Malaviya National Institute of Technology Jaipur
E-mail: 2022rec9034@mnit.ac.in; sjnanda.ece@mnit.ac.in; rpyadav.ece@mnit.ac.in

The edge server placement problem is a critical aspect of Mobile Edge Computing where data processing happens close to the user rather than relying on a distant cloud server. When the computing resources are kept much closer to the user then the response time is reduced to a great extent and also the data traffic congestion at the cloud is also minimized. In today's 5G networks majorly all the applications require no or very less delay and strong data connection as in the case of medical facilities, automated vehicles, or smart traffic the response time needs to be very fast. In this manuscript, the placement of servers is optimized to reduce energy consumption while maximizing user experience. This issue is designed as a single-objective optimization problem and a constrained heuristic optimization algorithm is designed to find the optimal solution. The algorithm is evaluated on real dataset from Shanghai Telecom and the results show our algorithm can reduce energy consumption to a significant amount as compared to other algorithms.

Contributory Sessions 8

Session SA.08.H3.X

AI IN COMMUNICATION SYSTEM DESIGN

294

Multi-User Multiple-Input Multiple-Output (MIMO) Communications Through Binary Flower Pollination Algorithm-Based User and Antenna Scheduling

Jyoti Mohanty[1*], Akhilesh Kumar[2], Suvrakanta Pal[3], Sabyasachi Patra[4], Suchismita Sathpathy[4] & Prabina Pattanayak[5]

[1]*Siksha O Anusandhan University*
[2]*Sandip University*
[3]*TITE, Bhubaneswar*
[4]*C V Raman Polytechnic*
[5]*NIT Silchar*

E-mail: jyotimohanty@soa.ac.in; akhilesh21_rs@ece.nits.ac.in; pal.bapu87@gmail.com; sabyasachi.patra@cvrp.edu.in; suchismita@cvrp.edu.in; prabina.pattanayak@ieee.org

In this paper, we take a look at multi-user multiple input multiple-output(MU-MIMO) communication system that uses the user scheduling mechanisms. When looking to boost system throughput and spectral efficiency, networks often use MU-MIMO communication technologies. Users are chosen at the base station (BS) according to the signal-to-interference-plus noise ratio (SINR). A base station with M transmits antennas will choose the M best antennas on the receiving side and send data in separate, parallel streams to those antennas. The combined user and receive antenna scheduling procedure faces a massive search space as the number of active users and receive antennas increases. Within the time constraints of wireless transmission, it is very challenging to explore the full search space. Methods for user and receiving antenna joint scheduling based on low complexity are presented in this study. The results demonstrate that a scheduling method based on low complexity may achieve system throughput that is comparable to an exhaustive search algorithm's (scanning the whole search space). The number of complex multiplications and additions that are needed by these low complexity approaches is much lower than that of an exhaustive search algorithm, which is an interesting observation that can be made from this observation.

323

Lower Order Polynomial-based Trajectory Generation for Mobile Robots

Snehil Sanyal*, Shantanu Choudhury & Shubhashis Sanyal

NIT Raipur

E-mail: ssanyal.phd2024.ece@nitrr.ac.in; shantanuchoudhury40@gmail.com; ssanyal.mech@nitrr.ac.in*

Efficient trajectory generation is critical for motion planning in robotics and autonomous systems, balancing path smoothness and computational cost. The present work presents a lower-order, 4-3-4 polynomial trajectory generation method optimized for better computational efficiency. The proposed method is further compared with RRT* and higher-order polynomial methods for smoothness of trajectory, computational efficiency, and optimization capability. The code is open-source and available at: https://github.com/shantanuc1205/4-3-4-Polynomial-Trajectory/

341

Optical Cryptosystem-based Multimodal Biometric Security

Gaurav Verma

LNMIIT

E-mail: gaurav.verma@lnmiit.ac.in

Multimodal biometric systems offer several benefits over unimodal systems, including broader population coverage, enhanced recognition accuracy, and reduced error rates. However, multimodal systems must preserve biometric data's integrity, security, and privacy since they include more than one biometric attribute linked to each user. This paper uses optical encryption to present an efficient framework for securing multimodal biometrics such as fingerprints, face, and iris scans. In the proposed scheme, we integrate the discrete wavelet transform (DWT) and the phase truncated Fourier transform (PTFT) to increase security and maintain the privacy of multimodal biometrics throughout the process. The involvement of the DWT allows fusion analysis to combine multimodal biometrics, and the PTFT process offers an optical encryption strategy. This approach enhances the security of multimodal biometric modality by ensuring high robustness against unauthorized attempts and attacks. Computer simulation shows the feasibility and efficacy of the idea.

416

Characterization of an Encoder-only Model on Heterogeneous Systems

Devashree Tripathy* & Soumya P. Dash

IIT Bhubaneswar

E-mail: devashreetripathy@iitbbs.ac.in; spdash@iitbbs.ac.in

The rapid adoption of large language models (LLMs) is driving an increase in inference demands, resulting in significant energy consumption in data centres due to high computing and memory requirements. We quantify performance, analyse the inference pipeline to identify opportunities for energy optimization, and outline the requirements for a comprehensive performance analysis framework. We examine the impact of tuneable parameters, as well as various NLP tasks such as text classification and question answering, on latency and energy consumption. This analysis aims to enable the design of optimized inference systems without compromising performance. This paper focuses on the BERT model as it is commonly utilized in inference-heavy tasks across various AI applications, preferred for its quick and efficient response. By analysing its computation and idle period, we can develop strategies to minimize power wastage and optimize resource utilization. Our findings focus on the idleness profiling of BERT model, which eventually translates to GPU under-utilization. We profile different compute-intensive modules of this model and analyse its energy behaviour with respect to the frequency scaling power management technique.

567

Optimizing Dual-Function Radar and Communication: A Performance Study of FMCW, OFDM, and OTFS

Tasmin Mishra*, Debalina Ghosh & Prasant Kumar Sahu

IIT, Bhubaneswar

E-mail: a24ec09011@iitbbs.ac.in; deghosh@iitbbs.ac.in; pks@iitbbs.ac.in

Orthogonal time frequency space (OTFS) modulation is a very competent technology for data transmission in high mobility scenarios. It has become very prevalent in recent years due to enhanced channel representation, full diversity gain, improved spectral efficiency, and robust channel estimation. In this paper, integrated radar sensing and communication in high mobility scenarios is demonstrated using OTFS, OFDM (orthogonal frequency division multiplexing), and FMCW (frequency modulated continuous wave) waveforms. For comparative analysis, three independent targets have been adopted, and the range and velocity are estimated using different estimation algorithms. FMCW modulation technique proves to be very efficient performance for noiseless scenarios, however, in noisy and high mobility environment OTFS provides better estimation, enabling wider ranges of doppler frequency estimation. At the same time, the waveforms are analysed to understand their performance in multi-user communication scenarios.

587

MI-STAR: Mitigation of Self-Interference in Simultaneous Transmit and Receive Array via Evolutionary Optimization

Amisha Verma* & Debalina Ghosh

IIT Bhubaneswar

E-mail: as74@iitbbs.ac.in; deghosh@iitbbs.ac.in

Simultaneous Transmit and Receive (STAR) technology is a promising alternative to conventional communication systems, offering improved spectral efficiency and capacity. However, self-interference (SI) remains a major challenge. This paper proposes a shared aperture array design optimized using Genetic Algorithm (GA) to minimize SI ensuring a balanced distribution of transmit and receive elements. A 10 × 10 antenna array with two coaxial ports per unit cell is used for validation, incorporating two-unit cell designs: a square patch antenna for higher frequencies and a crossbow-shaped patch for lower frequencies. GA-based optimization achieves substantial SI reduction, with -4.005 dB for the square patch and -8.608 dB for the crossbow structure. The results demonstrate that aperture optimization effectively enhances isolation, making STAR systems more effective for wireless communication, radar, and spectrum sharing applications.

596

SFDR Improvement in DDS Using Memory Compression Techniques

JS Mangat

DLRL, DRDO, Hyderabad
*E-mail: js.mangat.dlrl@gov.in**

Analysis and comparison of different memory compression techniques for Direct Digital Synthesizer (DDS) is presented in this paper. The important parameters of DDS are frequency resolution, number of frequency spots and frequency step size. These parameters are mainly dependant on the minimum phase increment of accumulator. Any improvement in these parameters requires width of the accumulator register to increase. As size of lookup memory grows exponentially with width of the accumulator output and hardware limitations restrict the size of memory which can be allocated for Lookup Table (LUT). To overcome this, accumulator output is truncated to match LUT size. Truncation, however, results in generation of spurious signal which reduces SFDR. There are some techniques available to compress this memory requirement. In this paper different compression techniques are compared for their performance. In this paper we will first be briefly discussing causes of spurious signals in the DDS circuitry and then we will be analysing various techniques/algorithms employed to minimize these causes to obtain cleaner output in DDS and improve its SFDR.

Contributory Sessions 9

Session SA.09.H4.X

ADVANCEMENTS IN HEALTHCARE USING AI

9

Verification of Himalayan Meditative State Using EEG Rhythms

Pankaj Sahu

Dr B R Ambedkar National Institute of Technology Jalandhar
*E-mail: pankajiiscb@gmail.com**

In the present work, we examine whether hypothesised variations among meditators and non-meditators in their subjective meditation experiences are mirrored in their EEG spectrum. To achieve this, we considered recorded EEGs of 12 novices and 12 experts in Himalayan meditation. Occipital and pre-frontal theta and alpha rhythms are essential for classifying the brain's meditative state. The multi-level (3-level) Discrete Wavelet Transform (MDWT) is performed with db2 wavelet to separate the theta and alpha rhythms in the same wavelet coefficient matric from the pre-frontal and occipital electrodes groups, where each group comprises three electrodes. The statistical features are computed from the theta and alpha coefficient matrices. Finally, the solo and ensembled machine learning classifiers are applied to classify the 280 features of occipital and pre-frontal theta and alpha rhythms. The research outcome is that the pre-frontal and occipital theta rhythms are inversely proportional to the pre-frontal and occipital alpha rhythms. The proposed model has achieved 100% accuracy for the ensembled Ada Boost Classifier, with the Random Forest Classifier achieving 83.33% accuracy. The MDWT, coupled feature extraction for the pre-frontal and occipital electrode groups with three EEG electrodes, effectively identifies the meditative state.

63

A Two-Stage Anomaly Detection Framework for Improved Healthcare Using Support Vector Machines and Regression Models

Shreea Bose, Venkata Saketh Dakuri & Chittaranjan Hota*

BITS-Pilani, Hyderabad Campus
E-mail: p20240026@hyderabad.bits-pilani.ac.in; f20220056@hyderabad.bits-pilani.ac.in; hota@hyderabad.bits-pilani.ac.in

This paper presents an effective two-stage anomaly detection technique for wireless body area networks (WBANs)-based personalized healthcare monitoring. Our method uses SMO regression (Sequential Minimal Optimization) for more in-depth contextual analysis and Support Vector Machines (SVM) for initial anomaly categorization. Domain-specific thresholds and sophisticated feature engineering are utilized to accurately identify point anomalies brought on by sensor noise or malfunctions and contextual anomalies resulting from interdependent physiological feature deviations. The robustness of the model is improved by features including heart rate discrepancies, rolling mean, and rolling standard deviation. With 99.61% and 99.97% detection accuracy for point and contextual anomalies, respectively, the model attains an overall accuracy of 99.59%. These findings highlight the promise of scalable, customized anomaly detection systems for WBAN-driven, real-time healthcare, which could lead to prompt treatments and better patient outcomes.

313

Deep Neural Network with Dilated Convolutions and MRI Imaging for Multi-Class Staging Prediction of Alzheimer's Disease Progression

Manan Vangani[1], Rakesh Chandra Joshi[1*], Abhijay[1], Abhishek Sengupta[1], Vinay Kumar Pathak[2] & Malay Kishore Dutta[1]

[1]*Amity University, Noida, India*
[2]*Chhatrapati Shahu Ji Maharaj University*
E-mail: manan.vangani@s.amity.edu; rakeshchandraindia@gmail.com; abhiashmani@gmail.com; asengupta@amity.edu; vinay@vpathak.in; malaykishoredutta@gmail.com*

Alzheimer's Disease (AD) is a progressive neurodegenerative disorder that impairs cognitive function and memory, making early and accurate diagnosis essential for effective intervention. Traditional diagnostic approaches are labour-intensive, requiring expert assessment of MRI scans, which may delay early-stage detection. In this study, a deep learning-based framework has been proposed employing dilated convolutions integrated with lightweight convolutional neural network architecture to classify AD progression across six distinct stages: Cognitively Normal (CN), Subjective Memory Concerns (SMC), Early Mild Cognitive Impairment (EMCI), Late Mild Cognitive Impairment (LMCI), Mild Cognitive Impairment (MCI), and AD. Dilated convolutional layers are incorporated to enhance feature extraction by expanding the receptive field without increasing computational complexity, allowing the model to capture subtle morphological changes in brain structures such as the hippocampus and cortex. The proposed model achieves a classification accuracy of 91%, outperforming conventional CNN architectures used in AD diagnosis. Results show that dilated CNNs can discriminate between the successive AD stages and help with early-stage diagnosis and early-stage therapeutic interventions. Ablation studies that compare the optimizers and the number of neurons in the dense layer find the proposed method robust. This work describes the promise of deep learning for automating AD detection, cost reduction in diagnosis and clinical decision-making by the use of AI-integrated radiological appraisal.

316

A Hybrid Approach for Emotions Classification using Modified Intrinsic Time Scale Decomposition

Priyadarsini Samal* **& Mohammad Farukh Hashmi**

National Institute of Technology Warangal

E-mail: priyadarsini.samal18@gmail.com; mdfarukh@nitw.ac.in

Emotion recognition is a crucial area of research in brain-computer interactions (BCI). The swift progress of brain-computer interfaces (BCI) has established emotion identification through electroencephalogram (EEG) data as a prominent area of research in recent years. The complexity and diversity of emotional signals render EEG-based emotion recognition a challenging endeavour in pattern recognition. To address this issue, this study develops an automated classification model utilizing the PYNQ-Z2 platform for the recognition of human emotions. This model employs a modified intrinsic time scale decomposition to analyse the complicated EEG signal by breaking it down into numerous proper rotational components (PRC). The ideal PRC is subsequently chosen and input into the Convolutional Neural Network for automated feature extraction. The feature set is subsequently assessed using three distinct classifiers: Quadratic Discriminant Analysis, Multilayer Perceptron Network, and Ensemble Voting Classifier. The study utilized the DEAP dataset and achieved accuracies of 94.11%, 94.70%, and 94.76% for the valence, arousal, and dominance domains, respectively. The proposed method demonstrates superior performance compared to other existing approaches. Additionally, the model is executed on the PYNQ-Z2 board, demonstrating that the proposed technique is compatible with hardware.

385

Improved Brain Tumour Diagnosis through Feature Selection and Hybrid Ensemble Models

Karishma Afrin, Kirtimayee Bag, Bhumika Rani Das & Debendra Muduli*

CV Raman Global University

*E-mail: afrinkarishma70@gmail.com; nehakirtimayee1803@gmail.com; bhumikadas980@gmail.com; muduli.debendra@gmail.com**

Brain and central nervous system tumours comprise more than 120 different types, making them a complex and significant health concern around the world. In particular, brain tumours are among the leading causes of cancer-related mortality. Early and accurate diagnosis plays an important role in improving patient survival rates. Although traditional machine learning algorithms have been used for brain tumour classification, they often face limitations in handling the complexity and diversity of tumour types. To overcome these challenges, our study proposes an advanced approach utilizing feature selection and ensemble learning techniques for enhanced brain tumour detection. We employed the minimum redundancy-maximum relevance method for reducing noise and improving model performance. Following feature selection, an ensemble learning strategy was implemented using Artificial Neural Networks, Random Forest and Back Propagation Neural Networks classifiers, combined with a voting mechanism for final classification. This approach not only improves diagnostic accuracy but also contributes to better clinical decision-making and patient outcomes. This ensemble approach leverages the strengths of each model, achieving an impressive accuracy of 98.27 %, thus significantly improving classification reliability.

Contributory Sessions 10

Session SA.10.H1.X

MICROWAVE AND MMWAVE SENSORS IN ADVANCED APPLICATIONS

365

3D LiDAR-Based Online Monitoring of Sodium Combustion Aerosols Concentration via Regression Analysis

Avinash Acharya[1*], Muthukumar G[2], Snehalatha V[2], Krishan Kumar[3], Sanjay Kumar Das[2] & B K Sreedhar[2]

[1]*Indira Gandhi Centre for Atomic Research, Homi Bhabha National Institute*
[2]*Indira Gandhi Centre for Atomic Research, Kalpakkam*
[3]*Safety Research Institute – AERB, Kalpakkam*
E-mail: aacharya@igcar.gov.in; muthukumarg@igcar.gov.in; sneha@igcar.gov.in; kgoyat55@igcar.gov.in; skd@igcar.gov.in; bksd@igcar.gov.in

Sodium is used as coolant in sodium cooled fast breeder reactor (SFR). Sodium combustion aerosol detection and monitoring are key to safe management of sodium fire events in SFRs. A technique has been developed to monitor sodium combustion aerosol concentration in real-time using the Ouster 3D LiDAR (Light Detection and Ranging) module, eliminating the need for chemical methods. The module uses an array of laser light of 865 nm wavelength, which undergoes scattering in presence of sodium combustion aerosol particles. An experimental set up is designed to generate Sodium Combustion Aerosol Concentration (SCAC) with varying concentration to study 3D LIDAR data response. The SCAC, ranging from 50 mg/m3 to 2030 mg/m3 is collected from enclosed chamber at regular intervals using the passive sampling method and analysed via Ion Chromatography, with an accuracy of ±1 ppm. These concentration values are compared with acquired 3D LiDAR data. The LiDAR module generates signal photon, range, reflectivity, and IR data. Among these, signal photon and reflectivity points vary in proportion to SCAC. Two polynomial equations are derived for these two parameters from measured range of SCAC. The developed polynomial models are compared and found in close agreement (±5%). Details of developed methodologies and experiments are presented in this paper.

369

Novel Technique for UWB based Breathing Detection

Sandeep Acharya[1*], Richa Dham[1] & Sukanth A[2]

[1]*Infineon Semiconductor Technologies India, Pvt. Ltd.*
[2]*Vellore Institute of Technology*
E-mail: sandeep.acharya2@infineon.com; richa.dham@infineon.com; sukanth02a@gmail.com

In this paper we provide novel technique for detection of Respiration Rate using UWB-Radar. Respiration rate, also called as breathing rate is the number breaths a human takes per minute. It is characterized by the expansion and compression of the thoracic cavity relating to inhalation and exhalation, respectively. Almost any activity that even has a very diminutive movement corresponding to it can be tracked using an Ultra-Wide Band Wave. Respiration Rate very often finds itself being seen as a neglected and overlooked Vital Sign mostly because of the lack of technological improvements and the need for contact-based analysis for a considerably long amount of time to arrive at a precise measurement that can be put to effective use. Especially in the post-covid era, there are a lot of medical research going on to provide an effective, efficient and an accurate contactless vital sign monitoring mechanism and to fuel the development of technologies supporting it. The proposed technique inherently and implicitly adopts the fundamental techniques like range-doppler map and phase tracking which are very effectively specific to UWB-RADAR when compared to other traditional RADAR systems which most of the already available techniques failed to advantageously utilize. Our novel solution aims to support this research by bridging this gap mentioned earlier in respiration rate measurement by providing an effective contactless way to calculate respiration rate using the UWB-RADAR.

389

Microwave Resonator Based Blood Glucose Sensor

Abhishek Kumar* & Amarjit Kumar

NIT Warangal
E-mail: ak23ecm3s01@student.nitw.ac.in; amarjitk@nitw.ac.in

The pursuit of non-invasive glucose monitoring has remained a cornerstone of biomedical research for the past three decades. Despite advancements in sensor technology, existing solutions often fall short in terms of accuracy, circuit complexity, and real-time applicability. Consequently, the demand for a cost-effective, pain-free, and reliable monitoring system persists. Current research trends focus on microwave and electromagnetic sensing methods in the ISM (Industrial, Scientific, and Medical) radio frequency bands, exploiting changes in blood and interstitial fluid dielectric properties induced by glucose variations. Among these, microstrip ring resonators (MRRs) are gaining prominence due to their compact structure, high sensitivity, and planar circuit compatibility. This study presents an enhanced coupling MRR designed to detect glucose concentration changes in aqueous solutions within biologically relevant ranges. Operating at 1.25 GHz, the sensor utilizes its transmission coefficient (S21) to monitor frequency shifts caused by glucose-induced dielectric variations. The resonator achieves a sensitivity of 3 MHz/100 mg/dL with a quality factor (Q) of 40, ensuring precise detection at low concentrations. Performance metrics reveal a voltage standing wave ratio (VSWR) of 1.1012, indicating optimal impedance matching and minimal signal reflection. Additionally, the return loss of-27 dB corresponds to an energy transmission efficiency of 99.97%. The optimized coupling periphery enhances electromagnetic field interaction within the resonator's core, improving coupling capacitance (Cg) and insertion loss. Comparative analysis highlights superior performance over conventional resonators in sensitivity, Q-factor, and power efficiency, establishing its potential for non-invasive glucose monitoring.

501

Wi-Fi CSI based Human Activity Monitoring using Machine Learning and Deep Learning algorithms

Ranjit Kumar*, Amit Sarkar & Debalina Ghosh

Indian Institute of Technology Bhubaneswar

E-mail: 20ec06020@iitbbs.ac.in; as58@iitbbs.ac.in; deghosh@iitbbs.ac.in

Monitoring of human activities is a broad field of research. Although there is sensor and vision-based solution is available, however this solution suffers from major limitations. like security and privacy concern as well as relatively expensive components. As a result, in order to alleviate or eliminate these limitations, Radio signal like Wi-Fi is best suitable solution. In order to mitigate above concern, most of the research work are done with the received signal strength indicator (RSSI) methodology. However, accuracy is the one of the major concerns in RSSI. So now a day channel state information (CSI) methodology is one of the popular research areas. However, extracting the

Wi-Fi channel state information (CSI) need expensive hardware components like USRP SDR or Network Interface Card (NIC) with PC that make this CSI extraction method too expensive. In my research work I reduce the cost factor by using one of the cheapest hardware called ESP32, through which I extracted the channel state information of received signal. Based on the channel state information, I detected the human activity like walk, sitting and falling inside a room as well as behind the wall. Since 2.4 GHz Wi-Fi signal can easily penetrate through wall. After extracting the channel state information (CSI), I apply the signal processing technique like Wavelet Transformation, Principal component analysis (PCA) in order of reduction of noise, signal transformation and features extraction. Then I proposed machine learning algorithm like K-Nearest Neighbour (KNN), Support vector machine (SVM), Decision Tree, Random Forest, and Recurrent Neural Network (RNN) for classification of different activity.

530

Bandwidth Enhancement of Slotted Hexagonal Patch Antenna for 6G Ultra-Fast Data Transfer and Brain Computer Interface

Uppala Rao*, Rajesh Kommalapati, Surendra Kumar Painam, Suneel Miriyala, Chandrasekhar Rao Jetti & Sanjay Bhargav Sakhinala

Bapatla Engineering College
E-mail: usrao75@gmail.com; rajeshkanna2050@gmail.com; mailtosurendra@gmail.com; suneel.miriyala@gmail.com; jettychandu@gmail.com; sanjaysakhinala@gmail.com

This study carefully designs and analyses a Slotted Hexagonal microstrip patch antenna operating at 28 GHz, 44.79 GHz, and 54.9 GHz to meet the requirements of 6G communications. A thorough evaluation is conducted on key antenna parameters, including gain, bandwidth, S-parameters, azimuthal and elevation half-power beamwidth, and directivity efficiency. The antenna is constructed using RT Duroid 5880, a high-performance substrate with a 1.6 mm thickness, renowned for its exceptional properties in high-frequency applications. Precisely engineered, the antenna resonates at 28 GHz, 44.79 GHz, and 54.9 GHz, aligning with the evolving needs of next generation communication systems. Among its notable performance characteristics, the antenna achieves directional gains of 4.92 dBi, 7.2 dBi & 5.88dBi demonstrating efficient radiation focusing. Additionally, its outstanding return losses of 24.68 dB, 26.06 dB, and 30.432 dB indicate superior impedance matching. Furthermore, the antenna features impressive bandwidths of 5.94 GHz and 18.34 GHz, enabling high-capacity data transfer and enhanced network performance in the millimeter-wave spectrum. To achieve optimal performance characteristics, CST simulations are employed to precisely refine the antenna's dimensions, substrate properties, and feed configuration.

553

Microwave Sensing for Bone Density Evaluation: A Time-Frequency Perspective

Nishtha Shrivastava*, Debalina Ghosh & Prasant Kumar Sahu

IIT Bhubaneswar

E-mail: a23ec09006@iitbbs.ac.in; deghosh@iitbbs.ac.in; pks@iitbbs.ac.in

This paper elaborates a method for evaluating bone densities using microwave signals. To extract meaningful features from bone signals, a variety of time-frequency domain transforms are used, including the Short-Time Fourier Transform (STFT), Wavelet Transform, Stockwell Transform, Wigner-Ville Distribution, Pseudo-Smoothed Wigner-Ville Distribution, and three Cohen Class distributions. Four key features—energy, entropy, centroid, and cross-term are analysed for four samples of varying bone densities. Moreover, the Bland Altman plot is also analysed to see the deviation of unhealthy bone type from healthy bone. The proposed method demonstrates the effectiveness of microwave systems in analysing bone health.

Contributory Sessions 11

Session SA.11.H2.X

COMMUNICATION SYSTEM DESIGN AND OPTIMISATION

204

A Novel Framework for Adaptive Jamming with Dynamic Null Zones for Securing Tactical Communications

Jyoti Ranjan Satapathy

Indian Army

E-mail: jyoti.ranjan1365@gmail.com

In modern battlefields, jammers play a crucial role but affect both friendly and enemy transmitters due to lack of selectivity. This paper presents an adaptive jamming system that selectively disrupts enemy networks. It protects friendly transmitters by creating a 500-meter null zone around them. The system operates in the 30-89 MHz military band with three 500 W jammers strategically placed for optimal coverage. The solution uses Federated Learning (FL) for Radio fingerprinting for accurate identification, Location Fixing, Dynamic adjustment of Phase and power for creation of the secure null zone. This work covers the theoretical framework, simulation methodology and phase and power control algorithms to cater for different environment and terrain conditions. The system safeguards friendly communication while jamming target areas which got significant tactical and strategic implications.

267

High Code Rate 16-QAM Golay Complementary Set Construction for Communication Systems

Aditya Prakash[1*], Palash Sarkar[2], Nishant Kumar[3] & Prashant Kumar Srivastava[1]

[1]*Indian Institute of Technology Patna*
[2]*Department of Informatic, University of Bergen*
[3]*Department of Mathematics, GLA University, Mathura*
E-mail: aditya_2121ma15@iitp.ac.in; palash.sarkar@uib.no; nishant.kumar1@gla.ac.in; pksri@iitp.ac.in

This paper presents two direct constructions of 16-QAM Golay complementary sets (GCS) employing generalized Boolean functions (GBFs) aimed at enhancing communication system performance. The initial method facilitates the generation of 16-QAM GCS with a set size of 2^{k + 1} and the constituent sequence of lengths in the form of 2^{m−1} + 2^t, offering advantages in terms of set size. The second methodology enhances the codebook size by utilizing GBFs, resulting in 16-QAM GCS with a sequence length of 2^m. In comparison to existing sequences, new sequences have a bigger family size, which leads to a higher relevant code rate which is beneficial for efficient data transmission. Notably, the construction demonstrates PMEPRs significantly lower than 4, i.e., the set size of GCS. This advancement holds promise for enhancing the OFDM system's efficiency and performance by exploiting the advantageous properties of Golay complementary sequences.

305

UAV-Assisted IRS-SIMO System in 6G Communication

Nikita Goel*, Pankaj Kumar, Pranav Chavare & Manoj Tolani

Manipal Institute of Technology Manipal
E-mail: nikita.goel@manipal.edu; pankaj.kumar@manipal.edu; pranav.chavare@manipal.edu; manoj.tolani@manipal.edu

This paper integrates UAV with IRS to enhance the reliability of SIMO-assisted D2D users in B5G and 6G, particularly in scenarios without operational base stations. This proposal is useful for improving the reliability in infrastructure less scenarios in areas where the base station is inoperable. The mathematical framework for the UAV-assisted-IRS-SIMO-D2D (UA-IRS-SIMO-D2D) communication system has been developed and derived the expression of spectral efficiency and closed-form expression of an outage probability as a performance metrics. The theoretical results are validated by using MATLAB and compared it with the existing cutting-edge systems. The results highlight the significant improvements offered by drone assisted D2D communication system. The results indicate that dynamic positioning of drone allows for enhanced SNR, increased data transfer rates, with minimized signal propagation and optimized energy utilization. This research underscores the potential of drone-assisted IRS to address shortcomings of traditional wireless systems.

384

Delay Domain Reuse for Multi-cell OTFS

Amit Kumar Pathak[1], Venkatesh Khammammetti[2] & Saif Khan Mohammed[1*]

[1]*Department of Electrical Engineering, Indian Institute of Technology Delhi*
[2]*Department of Electrical and Computer Engineering, Duke University*
E-mail: amit.kumar.pathak@ee.iitd.ac.in; venkatesh.khammammetti@gmail.com; saifkmohammed@gmail.com

We consider the problem of severe inter-cell interference (ICI) to a cell-edge user terminal (UT) in an Orthogonal Time Frequency Space (OTFS) based multi-cell system where all cells fully reuse the delay-Doppler (DD) domain resource elements (RE). We propose a novel method for delay domain reuse of REs, which is shown to achieve low SNR spectral efficiency (SE) performance close to that achieved in a single-cell system with no ICI. The low SNR SE achieved with the proposed reuse method is better than that achieved with a full reuse multi-cell system.

418

Design and Implementation of Molecular Communication (MC) Test Bed

Praneeth Mangala*, Anamika Singh, Vedant Lokhande & Prabhat Sharma

Visvesvaraya National Institute of Technology

E-mail: bt21ece054@students.vnit.ac.in; Anamikasingh@ece.vnit.ac.in; bt21ece083@students.vnit.ac.in; prabhatsharma@ece.vnit.ac.in

This project introduces a 3-bit (8-node) Molecular Communication (MC) setup, where liquid acts as the medium for binary data transmission. Normal water represents bit '0', and a bleaching solution represents bit '1'. Water serves as the carrier, and the bleaching solution is the message signal, with modulation controlling water flow to encode the data from transmitter to receiver. The system consists of a transmitter and a receiver, where water motors pump normal water and bleaching solution, with solenoid valves controlled by a microcontroller encoding bit '1' as bleach and bit '0' as water. A pH sensor at the receiver detects chlorine for bit '1', while its absence indicates bit '0', with an Arduino decoding the data. Solenoid valves act as nodes, sequentially modulated to create an 8-node network, ensuring accurate transmission and minimal errors. In a 3-bit (8-node) molecular communication system, data transmission is handled using an n-bit algorithm that dynamically controls relay activation. The system converts a 3-bit binary input (e.g., "101") into its decimal equivalent using the formula: Decimal Equivalent = *b*^0.2^0 + *b*^1.2^1 + *b*^2.2^2 where *bi* represents the *i*-th bit (0 or 1) in the binary input. The receiver validates the input and activates the corresponding relay using: Relay Pin=relayPins[Decimal Equivalent] to transmit molecular signals, accordingly, ensuring scalability and efficient relay management. The methodology flowchart mentioned in the paper outlines the step-by-step process of molecular communication, starting from binary data encoding using solenoid valves at the transmitter to decoding via a pH sensor at the receiver. It visualizes relay activation, data transmission, and validation, ensuring efficient and accurate signal processing. This project's novelty lies in using liquid as a communication medium, with solenoid valves modulating water flow to transmit binary data. The system's 8-node structure ensures reliable data transfer and decoding, making it suitable for specialized environments, such as biomedical and underwater communication, where traditional methods may not work. The Molecular Communication (MC) testbed, using liquid-based signalling with chlorine to encode binary data, offers a novel solution for environments where traditional methods like RF, optical, and electromagnetic waves face challenges. In biomedical fields, MC can revolutionize targeted drug delivery by enabling precise communication with nanodevices inside the human body, triggering localized treatment. Underwater communication, where RF fails, can also benefit, allowing submarines or autonomous underwater vehicles (AUVs) to communicate effectively via molecular signals. Similarly, in industrial chemical process monitoring, MC provides reliable data transmission in harsh environments, such as oil refineries or wastewater treatment, where conventional methods struggle. Additionally, environmental monitoring and agriculture can leverage MC for real-time soil or ecosystem assessments. Its scalability, adaptability, and ability to overcome traditional signal limitations make MC a promising technology for diverse real-world applications, from medical breakthroughs to advanced industrial and environmental solutions.

430

PIC based Implementation of DPS-QKD Transceiver

Ashay Atkar*, Harsh Saraf, Vinod Kumar, Anamika Singh & Prabhat Sharma

Visvesvaraya National Institute of Technology
E-mail: bt21ece060@students.vnit.ac.in; bt21ece100@students.vnit.ac.in; bt21ece031@students.vnit.ac.in; anamikasingh@ece.vnit.ac.in; prabhatsharma@ece.vnit.ac.in

Photonic integrated circuits (PICs) offer significant advantages, including miniaturization, enhanced stability, mass producibility, and high-speed operation. However, limited work has been done on the use of PICs for Quantum Key Distribution (QKD), more specifically, for Differential Phase Shift QKD (DPS-QKD). DPS-QKD offers simplicity, high key generation efficiency, and resilience against photon-number splitting attacks, making it a promising quantum cryptography protocol. This work presents the working model of DPS-QKD using Lumerical interconnect, which is optimized for PIC design. Further we have analytically analysed the performance of this PIC design in terms of the Quantum bit error rate (QBER) and secured key rate it can provide. Our findings indicate that a secured DPS - QKD link can be achieved for a 200 km fiber length with a QBER of 3.07% and a secured key rate of approximately 1.8 Kbps.

451

Enhancing Security through Correlation and Integration of Multi-Source Network Logs

Shivaram GV*, Santhosh H, Vinith SB, S Sri Harshith & Aslam B Nandyal

Alva's Institute of Engineering and Technology

E-mail: 4al21cs142@gmail.com; santhusanju903@gmail.com; vinithsb999@gmail.com; sriharshith44@gmail.com; aslamnandyal@gmail.com

In the ever-evolving digital landscape, securing networks against advanced cyber threats requires innovative approaches beyond traditional security systems. This article presents a comprehensive framework for enhancing network security through the integration of multi-source log data and machine learning techniques. The proposed system aggregates and normalizes logs from various sources such as firewalls, servers, routers, and endpoint devices to ensure seamless analysis. Using advanced machine learning models, including clustering algorithms such as K-Means and DBSCAN and neural networks such as autoencoders, the system identifies anomalies indicative of potential threats. An event correlation module further enhances threat detection by linking related activities across multiple sources, reducing false positives, and identifying coordinated attack scenarios. The integration of real-time threat intelligence feeds ensures adaptability to emerging attack vectors and vulnerabilities. In addition, an artificial intelligence-powered automated response system mitigates the identified risks, providing timely and effective countermeasures. The solution is scalable, efficient, and capable of real-time monitoring, which makes it suitable for organizations of varying sizes. By combining machine learning, event correlation, and threat intelligence, this approach addresses modern cybersecurity challenges, offering robust threat detection and mitigation capabilities. Key features include anomaly detection, real-time monitoring, and AI-driven automated response, which together form a resilient defence against sophisticated cyberattacks.

Contributory Sessions 12

Session SA.12.H3.X

SENSING AND TRACKING APPLICATIONS

187

Adaptive Histogram Equalization with Fine-Tuned DNN-YOLOv9 Technique for Vehicle Detection in Adverse Weather

Priya Ranjan, Ankita Jaiswal, Zeba Kausar, Devesh Kumar & Ram Chandra Barik*

C V Raman Global University, Bhubaneswar, Odisha, India
E-mail: ranjanpriya22236@gmail.com; ankitaariyaa11@gmail.com; zebakausar16@gmail.com; devesh2462003@gmail.com; ram.chandra@cgu-odisha.ac.in

Autonomous vehicles face significant challenges in adverse weather, where rain, fog, snow, and sandstorms reduce visibility and impair sensor performance. Cameras struggle with contrast and detail, lidars face backscatter issues, and radars encounter signal interference, collectively impacting perception and decision-making. This research proposes a solution combining advanced image enhancement techniques with state-of-the-art object detection models. Enhanced Super-Resolution Generative Adversarial Networks (ESRGAN) and Histogram Equalization improve image quality by enhancing contrast and detail in low-visibility scenarios. The YOLOv9 object detection model, known for its accuracy and real- time processing, ensures reliable detection in such conditions. Using the DAWN dataset, this methodology integrates image enhancement and robust detection to address weather-induced perception challenges. Results indicate improved object detection accuracy and reliability, enabling safer navigation in diverse environments. This study advances autonomous vehicle technology by mitigating weather-related limitations, supporting consistent and dependable operation across adverse and unpredictable conditions.

299

Fiber Modal Interferometric Networks for Enhanced Vehicle Identification and Traffic Flow Analysis

Arvind Maurya, Kalipada Chatterjee & Rajan Jha*

IIT Bhubaneswar

*E-mail: s21ph09011@iitbbs.ac.in; rkc11@iitbbs.ac.in; rjha@iitbbs.ac.in**

This paper describes a fiber optic interferometer-based system for real-time vehicle identification and characterization. The system consists of two in-line fiber interferometers embedded in road material to detect essential vehicle parameters such as speed, weight, axle count, and wheelbase. The proposed system recognizes a wide range of vehicles, including lightweight motorcycles to passenger buses, with high fidelity. With a sensitivity of 40 pm/100 kg, the system allows for exact weight estimation, and its real-time signal acquisition capabilities make it useful in traffic monitoring, vehicle categorization, and weight-in-motion applications. The technology is immune to electromagnetic interference with high stability and tolerance towards environmental changes.

325

Geo-tagging and Tracking for Detected Vehicles using UAV

Avani Vipat*, Parnika Das, Shubh Gupta & Vaishnavee Rathod

Sardar Patel Institute of Technology

E-mail: avani.vipat22@spit.ac.in; parnika.das22@spit.ac.in; shubh.gupta@spit.ac.in; vaishnavee.rathod@spit.ac.in

Vehicle tracking and geo-tagging have gained significant attention due to their applications in traffic monitoring, toll management, and urban planning. This research focuses on two components—vehicle tracking and geo-tagging/geo-registration—using drone footage from the SHIV–NATARAJ dataset, specifically at a toll plaza. A YOLO-based object detection and ByteTrack tracking system was implemented to detect and track vehicles across frames, assigning unique IDs for movement analysis. Separately, a homography transformation technique was used to overlay detected vehicles onto Google Earth, ensuring accurate latitude and longitude tagging. The methodology involved frame-by-frame vehicle detection, persistent ID assignment for tracking, preprocessing the dataset, developing a model for vehicle classification, and implementing a geo-registration system using UTM coordinate transformation. The results demonstrate reasonably accurate vehicle tracking with minimal ID switching and a precise alignment of geo-tagged vehicles with real-world coordinates. This study contributes to intelligent transportation systems by integrating computer vision-based tracking and geospatial mapping, offering insights into vehicle movement patterns and spatial positioning at toll plazas.

402

Robust Vehicle Detection in Aerial Imagery: Comparative Study of YOLO Variants

Diya Kulkarni[*], Shrutika Yeole, Soham Karmarkar, Shubh Gupta & Vaishnavee Rathod

Sardar Patel Institute of Technology

E-mail: diya.kulkarni22@spit.ac.in; shrutika.yeole22@spit.ac.in; soham.karmarkar22@spit.ac.in; shubh.gupta@spit.ac.in; vaishnavee.rathod@spit.ac.in

Vehicle detection from satellite images requires strong models to handle varied scales, low resolutions, and cluttered backgrounds. Conventional object detection methods underperformed in these environments. This paper analyses the performance comparison of YOLOv8, YOLOv11, and YOLOv12 for vehicle detection based on the SHIV-NATRAJ dataset. The pipeline is designed to be well-organized, comprising frame extraction, annotation, and YOLO format conversion. The models were trained with optimized hyperparameters and enhanced using data augmentation techniques. The performances are compared using essential measures such as precision, recall, F1-score, mAP@50, and fitness score to determine the best YOLO variant for vehicle detection in the aerial domain.

576

Unitary-ESPRIT for 2D-DOD and 2D-DOA Estimation with EMV Sensors in MIMO Radar

Srinivasarao Chintagunta[1*] & Karthick Subramaniam[2]

[1]*National Institute of Technology Calicut*

[2]*NNRG Institutions*

E-mail: srinivasarao@nitc.ac.in; karthick.s@ece.nnrg.edu.in

This paper presents the implementation of a unitary ESPRIT method for estimating the joint two-dimensional directions of departure (2D-DOD) and two-dimensional directions of arrival (2D-DOA) of coherent targets in multiple-input multiple-output (MIMO) radars equipped with uniform rectangular arrays of electromagnetic vector (EMV) sensors. The proposed method employs real-valued operations in the computation of the covariance matrix and its eigenvalues decomposition and can resolve up to 72 coherent targets without requiring additional pre-processing for smoothing. Monte Carlo simulations are conducted, and the results are compared with the Cramér-Rao bound to validate the method.

Contributory Sessions 13

Session SA.13.H4.X

COMMUNICATION AND SENSING ON IOT PLATFORMS

44

Integrating IOT Devices for Monitoring and Enhancing Working Conditions of Workers in India

Abhishek Ranjan, Tushar Pati Tripathi, Rohan Joshi, Aiyushi Srivastava, Rishita Jauhari, Sasikumar Periyasamy* & Maheswari S

VIT, Chennai

E-mail: abhishekranjansingh001@gmail.com; tushartripathi851@gmail.com; rohan155joshi@gmail.com; aiyushisrivastava20@gmail.com; jauharirishita@gmail.com; sasikumar.p@vit.ac.in; maheswari.s@vit.ac.in

In recent years, the Internet of Things (IoT) has emerged as a transformative technology with immense potential to revolutionize various industries, including the field of labour welfare. IoT serves and motivates health monitoring systems where sensors are either worn inside or outside the human body. The sensors collect the information of physical and logical conditions and the movements of the patient. This research paper explores the integration of IoT devices to record and monitor the working conditions of workers in India, with the aim of improving their overall well-being and productivity. These sensors are incorporated and interfaced with Arduino UNO board, ESP8266 for wireless data transmission to Thing Speak. The outcome of the result on Thingspeak is Data visualization. So that record of data can be stored and accessed over a period of time.

406

U-Net: A Binary Segmentation Architecture for Building Localization in Structural Health Monitoring using the xBD Dataset

Sampa Sahoo[1*], Sakshi Suwarna[2], Tanisha K[2], Sanchali Srivastava[2] & Khushi Raj Das[2]

1National Institute of Technology, Rourkela

2C.V. Raman Global University, Bhubaneswar

E-mail: sampaa2004@gmail.com; suwarnasakshi593@gmail.com; tanishastuti1127@gmail.com; sanchalisej134@gmail.com; khushirajdas259@gmail.com

Structural Health Monitoring(SHM) plays a vital role in disaster scenarios as they can be helpful in building localization, building damage assessment and classifying the extent of damage. SHM refers to the process of monitoring and tracking the condition of engineering structures over a period of time to detect damages, deterioration, or potential failures. Localization in such scenarios aids in identifying the buildings and generating ground truth. Traditional techniques of damage assessment are often manual and increases the labour cost due to onsite inspection. This paper aims to propose a methodology based on image segmentation which includes U-Net, a Deep Neural Network (DNN) based architecture with a ResNet-50 encoder and custom decoder. It is used to detect and localize buildings from the predicted masks and classify them as Building Detected (1) and No Building (0) based on the probability scores. The proposed approach has achieved a test accuracy of 0.9727 and an Intersection over Union (IoU) score of 0.6289 and a Dice Coefficient score of 0.7721 on all 11 disaster types present in the xBD dataset. The proposed model is further compared with other segmentation architectures like Fully Convolutional Networks (FCNs) and Attention U-Net based on IoU and Dice Coefficient scores. The proposed model shows better performance as compared to the existing models.

409

Internet of Things to Steer Point-of-Care Testing of Vital Sign Monitoring

Richa Khandelwal

Shri Ramdeobaba College of Engineering and Management, Nagpur
E-mail: khandelwalrr@rknec.edu

This research proposes a novel use of Internet of Things (IoT) technology for non-invasive patient vital sign monitoring. The measurement of parameters including oxygen saturation, heart rate, temperature, and Electrocardiogram (ECG) has traditionally been done using heavy-duty patient monitors. On the other hand, in the presented work, the portable device was developed which contains different sensors which are interfaced with a Wi-Fi enabled microcontroller development board to remotely sense these vital signs via the IoT. With this configuration, real-time data that is necessary for remote patient monitoring can be collected and easily accessed on portable devices. Our strategy enables timely point-of-care (PoC) diagnosis and therapy, hence reducing the risk of future health difficulties, by giving healthcare professionals quick access to vital patient data, such as important indicators of the patient's medical history and real-time changes in the patient's condition.

506

Enhanced GIFT Encryption with Chaotic Maps for Secure Data Transmission in LoRa-Based IoT Devices

Balakrishnan Ramalingam

Vellore Institute of Technology Chennai
E-mail: balakrishnan.r@vit.ac.in

LoRa-based IoT devices lack encryption, so secure data transmission is a crucial challenge. This work proposes a lightweight encryption scheme for LoRa-based IoT devices to secure data transfer, specifically for LoRa-based innovative metering applications. To address this, we use a combined chaotic map-based keystream generator to enhance the GIFT lightweight encryption algorithm. Using a nonlinear weighted sum approach, our approach combines a 1D chaotic system, such as Piecewise Linear Chaotic Map (PWLCM), and tent maps. This approach is computationally efficient with increased security, ensuring robust and lightweight data protection for LoRa-based smart devices' secure data transmission. The efficiency of our proposed system was validated using NIST randomness tests and execution time in various IoT-based embedded hardware platforms. The experimental results show that our proposed system passes all NIST randomness tests and takes low execution time.

569

A Novel Approach for an Enhanced Security System on IoT Devices

KARTHIK P*, PAVAN B V, CHETHAN B V & RUDRAGOWDA PATEEL

Cambridge Institute of Technology

E-mail: karthik.p@ieee.org; pavanbv.22ise@cambridge.edu.in; chethan.21ise@cambridge.edu.in; rudragowda.22ise@cambridge.edu.in

IoT devices increase the variety and complexity of security. Current security solutions typically fall short in a number of areas, especially when it comes to detecting and responding to attacks in real time. Through the interaction. Some examples are e-commerce, smart industrial management, smart grid, smart health, smart home, smart workplace, and e-governance. Every day, more and more devices get connected, which raises security risks and issues A comprehensive IoT security model is needed to support resource-based IoT gadgets and end security. We focus on the networks and IoT system applications in addition to the organizational approach towards IoT security, attack vectors, and security needs for IoT systems. We also provide a baseline for security implementation and recommend a security architecture to offer security-enabled IoT services through the application of this cybersecurity framework, the project aims to strengthen the security stance of IoT devices, reduce the attack surface, and foster user trust in IoT applications. By establishing best practices and guidelines, this framework serves as a foundational step toward creating a safer IoT environment, ultimately contributing to the broader goals of digital Privacy and security in and an interconnected world. Experimental results demonstrate the effectiveness of AI/ML techniques in decreasing down on false positives and enhancing response times, making IoT deployments more resilient to emerging security threats. This study highlights how AI/ML has the potential to completely security protocols in IoT, enabling safer and more reliable device integration across an assortment of fields.

577

Design and Development of a Novel Multi-functional ESP32-based Smart Device Featuring Optical Communication, Photometric Analysis, and Real-Time Data Processing

Ural Oinam

IIT Madras
E-mail: uraloinam@gmail.com

This paper presents the design, prototyping, and development of a smart multi-functional device that integrates photometry, data processing, optical communication, and real-time display capabilities. Powered by the ESP32 microcontroller, known for its dual-core architecture and 240 MHz processing speed, the device efficiently handles tasks such as real-time signal processing, data communication, and graphical visualization. A key feature is its photometric capability, where optical power is measured using a Light Dependent Resistor (LDR). While LDRs are generally unsuitable for optical power measurement due to their coarse sensitivity and non-linear response, these drawbacks are mitigated in this model through post-data collection and curve fitting, making it capable of high-end applications. Using black-dyed UV epoxy for protection from ambient light coupled with a slot for fluorescent fiber makes it capable for UV detection. For Data processing we propose a noble algorithm i.e., "Reference Line Conditional Averaging Filter," an improved version of the conventional "Run Length Averaging Filter," and the proposed filter enhances accuracy by reducing noise, smoothing signals, and enabling precise real-time data analysis. The device also supports optical communication, utilizing the ESP32's high clock speed and analog-to-digital converter to enable free-space communication within the device itself, facilitating visualization of optical data transfer. This integration of optical communication opens up future scope of remote free space optical data transfer between different modules or devices. With its comprehensive features, cost-effective design, and broad applicability, this smart device offers an adaptable and practical solution to various fields such as scientific research, industrial automation, and environmental monitoring.

595
React.js and Django REST Framework Architecture for IoT Applications

Shankar S

Centre for Development of Advanced (C-DAC), Thiruvananthapuram
E-mail: shankarss@cdac.in

The Internet of Things (IoT) has made rapid advancement in industries by enabling interconnected devices to collect, process, and manage real-time data effectively. Developing IoT applications requires scalable, efficient, and user-friendly architectures. This paper proposes a novel architecture that harnesses React.js, a dynamic JavaScript library for building responsive user interfaces, and Django REST Framework (DRF), a Python-based toolkit for creating RESTful APIs. The architecture supports real-time data visualization, secure data communication, and modular backend management. Technologies such as WebSockets and MQTT protocols are integrated to facilitate seamless real-time communication with IoT devices. This approach highlights the versatility and adaptability of the architecture, demonstrating its applicability across diverse IoT domains and use cases.

Contributory Sessions 14

Session SU.14.H1.X

METASURFACES FOR FIELD SHAPING AND SYNTHESIS

345

A Concave Circular Brick-Shaped Metamaterial-Inspired Superstrate Based High-Gain Microstrip Patch Antenna for Wideband Applications

Uday Kumar

BIT, Mesra

E-mail: ranjanuday333@gmail.com

In this manuscript, a highly effective metamaterial (MM) superstrate is presented, which is intended to significantly enhance the gain (G) of a partial ground wideband microstrip patch antenna (PG-WMPA) in the measured operating frequency band (OFB) of 3.8 GHz–10.65 GHz for satellite communications, 5G networks, and radar systems. A 3 × 3 array of the MM superstrate (MMS), sized 0.93λg×0.93λg, is formed using MM unit-cells with dimensions of 0.3λg × 0.3λg. Each unit cell incorporates an innovative integrated concave circular brick (ICCB) shaped along with horizontal and vertical rectangular slots. Using the proposed ICCB-shaped unit-cells, a 3×3 array of MMS is designed on an RT/Duroid 5880 substrate and positioned 14.5 mm above the PG-WMPA. The proposed MMS shows the negative intrinsic material properties in the wide frequency band from 5.8 GHz – 10.0 GHz. As a result of successfully implementing the proposed ICCB-shaped MMS, the measured average gain, Gav of the PG-WMPA increased from 4.49 dBi to 10.45 dBi and provided a bi-directional radiation pattern. Further, a prototype of the proposed PG-WMPA with ICCB-shaped MMS has been developed and measured to verify the genuineness of the simulated results.

388

Design of Chalcogenide Phase Change Metamaterial-based Microwave Absorber

Atchyut M

IIT Madras

E-mail: ee20d430@smail.iitm.ac.in

Active tuneable materials have emerged as a promising solution for achieving frequency and amplitude tunability in microwave absorbers. In this work, we have designed a metamaterial-based microwave absorber utilizing a phase change material (PCM) to enable dynamic control. The tunability is achieved by manipulating the phase state of the PCM, transitioning from the high-resistance amorphous state to the low-resistance crystalline state. Our design demonstrates a significant absorption amplitude contrast of approximately 93% at around 15GHz. The designed structure exhibits good angular stability, maintaining similar absorption contrast over a wide range of incident angles (0° to 40°). It also remains the same for TE and TM propagation modes, making it highly useful for practical applications.

400

Conformal Metasurface based Linear to Circular Polarization Converter for Dual Band Wi-Fi Applications

Karthik P B[1], Sai Sreenija Yallanki[2*], Rishi Mishra[2] & Harish A. R.[2]

1IIT Madras

2IIT KANPUR

E-mail: karthikpb@smail.iitm.ac.in; yssreenija22@iitk.ac.in; rishim@iitk.ac.in; arh@iitk.ac.in

A two-layer substrate based linear-to-circular polarization conversion (LCPC) conformal metasurface (CM) for two operational bands is proposed. The unit cell of the structure consists of two layers of substrate, separated by air gap, and with identical metallic layers of crossed dipoles on outer sides of both the layers of the substrate. The metasurface formed by this unit cell is tilted to 45° and thus wrapped onto a cylindrical surface to form a conformal structure. This is fed by a dipole antenna placed at the center of the cylindrical surface. Right-handed circularly polarized (RHCP) and left-handed circularly polarized (LHCP) beams are generated at 2.4 GHz and 5 GHz, respectively, when excited by the linearly polarized (LP) dipole antenna. The converter covers two operational bands for Wi-Fi communications with Axial Ratio (AR) as low as 2 dB.

408

A Frequency-Agile Filtering Metasurface Antenna with Single, Dual, and Wideband Reconfigurability

Sai Sharanya Vattam*, Amarjit Kumar, Sai Rohith Somaraju, Safiyya Ishaq Syed & Laranya Rane

National Institute of Technology Warangal
E-mail: vs21ecb0b63@student.nitw.ac.in; amarjitk@nitw.ac.in; ss21ecb0b58@student.nitw.ac.in; si21ecb0f21@student.nitw.ac.in; lm21ecb0f31@student.nitw.ac.in

This paper presents the design, analysis, and implementation of a reconfigurable filtering metasurface antenna (FMSA) that exhibits distinct single-band, dual-band, and wideband responses based on switching characteristic of pin diode. The antenna is designed to provide high frequency selectivity, enhanced out-of-band suppression, and high radiation efficiency for applications in wireless communication, radar systems, and cognitive radio. By integrating multi-folded U-shaped slots, defected ground structures (DGS), and coplanar parasitic patches, the antenna effectively suppresses unwanted frequency components while maintaining robust in-band radiation characteristics. The inclusion of PIN diodes in the design enables real-time frequency reconfigurability, allowing the antenna to operate in eight distinct frequency states by controlling the on and off states of different diode groups. This dynamic adaptability makes the antenna suitable for adaptive communication systems that require flexible frequency tuning and high spectral efficiency. The proposed FMSA exhibits single-band responses at 7 GHz, 7.2 GHz, 7.5 GHz, and 7.45 GHz, with bandwidths ranging from 0.03 GHz to 0.41 GHz. It also supports dual-band functionality at 6.4 GHz and 7.45 GHz, at 7.3 GHz and 7.5 GHz, demonstrating its ability to cater to diverse frequency requirements. Additionally, the antenna achieves a wideband response spanning 0.8 GHz from 6.8 GHz to 7.6 GHz, enhancing its potential use in broadband applications. The compact structure, with overall dimensions of $33 \times 33 \times 1.778$ mm^3, ensures seamless integration into modern electronic systems while maintaining high performance. The design methodology involves the modification of metasurface elements to improve the radiation characteristics and impedance matching properties, ensuring minimal insertion loss and improved gain across all frequency bands. The defected ground structure (DGS) is used for perturbing the electric field distribution, leading to enhanced frequency selectivity and sharper roll-off characteristics. The combination of U-shaped slots and coplanar parasitic patches further improves the antenna's filtering characteristics by generating broadside radiation nulls and additional resonance points. These features contribute to a high gain, wideband operation, and strong interference suppression, making the antenna a suitable candidate for next-generation wireless systems. The reconfigurability mechanism is achieved through PIN diode switching, wherein the diodes alter the effective electrical length and current distribution of the radiating structure. The selected MACOM MA4AGP907 PIN diodes offer low forward resistance, high reverse resistance, minimal capacitance, and fast switching speed, ensuring seamless transitions between frequency states. Three independent bias voltages control the diode groups, leading to the formation of eight distinct frequency states with varying band responses. The on state of a diode establishes a continuous current path, lowering the resonant frequency, while the off state disconnects patches, shifting the resonance to higher frequencies. This enables fine-grained control over the antenna's frequency response, allowing real-time adaptation to changing frequency environments. Comprehensive simulations and experimental validation confirm the effectiveness of the proposed FMSA in achieving superior filtering and reconfigurability. The return loss (S11) characteristics across different diode configurations indicate excellent impedance matching and strong frequency selectivity. The realized gain measurements further demonstrate stable radiation patterns with consistent gain performance across all operating bands. The ability to switch between narrow-band, dual-band, and wideband responses without requiring hardware modifications makes the antenna highly versatile for applications such as 5G communications, satellite systems, radar networks, and cognitive radio.

542

Capacitive-loaded Ultra-thin Polarization insensitive Dual-band Metamaterial Absorber

Punyatoya Routray

IIT BHUBANESWAR

E-mail: pr30@iitbbs.ac.in

This article presents a capacitive-loaded, low-cost, ultra-thin, dual-band, wide oblique, and polarization angle insensitive metamaterial absorber for X-band applications. The structure is designed on the concept of both putting more resonating elements in a unit cell and capacitive loading, which is applied by inserting the slots in the design. It consists of two resonating elements named as outer and inner resonating structures. The outer resonating structure is formed by a slot-loaded outer ring and outer split ring, whereas the inner resonating structure is formed by slot-loaded inner ring, inner split ring, and a crossed rod. The structure provides two resonating frequencies at 9.29 and 10.36 GHz with absorptivity of more than 90% and also provides a bandwidth of 500 MHz, and 500 MHz at center frequencies. The design is polarization insensitive in nature and provides high absorptivity up to 45° oblique incidence angle. The impedance matching theory and surface current density plots are used for the understanding of the mechanism of absorptivity. Index Terms—Capacitive-loaded, metamaterial absorber, oblique incidence, polarization insensitive, slots, X-band.

543

Reconfigurable Rasorber with Wide Absorption Band and Switchable Refection/Transmission Band for 5G and Bluetooth Applications

Punyatoya Routray

IIT BHUBANESWAR

E-mail: pr30@iitbbs.ac.in

This article proposes a reconfigurable rasorber consisting of a top frequency selective surface (FSS) loaded with resistors and a bottom FSS loaded with a pin diode for switching purposes. The top and bottom FSS are separated by an air gap, and the top FSS with a bottom ground plane acts as a wide absorber having an absorption band in the mid-band of sub-6 GHz for 5G applications. The primary advantage of the proposed rasorber is its ability to adjust the reflection/transmission band, transitioning from perfect reflection to bandpass response at the frequency of 2.4 GHz by switching the state of the PIN diode. For both the ON and OFF switching conditions of the rasorber, it provides a wide absorption band, approximately from 3 to 6.8 GHz with absorptivity of 90%.

566

Design and Optimization of a 5G Conformal Microwave Absorber for the n78 Band

Priyanshu Mathur*, Munasa Yuvaraju, Punyatoya Routray & Debalina Ghosh

Indian Institute of Technology Bhubaneswar

E-mail: 21ec01044@iitbbs.ac.in; my12@iitbbs.ac.in; punyatoya10@gmail.com; deghosh@iitbbs.ac.in

Rapid deployment of 5G networks has raised concerns about electromagnetic interference (EMI) and energy dissipation, necessitating the development of high-performance microwave absorbers. This study presents a conformal microwave absorber optimized for the n78 band (3.3–3.8 GHz), specifically targeting 3.6 GHz, a critical frequency for 5G applications. The proposed absorber is designed using CST Studio software and employs a low-cost, flexible substrate, enabling integration into curved surfaces while maintaining high absorption efficiency. Simulation results indicate an absorption rate exceeding 98% at 3.6 GHz, with broadband operation spanning 3.4–3.8 GHz. A detailed parametric analysis of substrate material, thickness variation, and periodic unit cell configurations was conducted to maximize performance. The findings suggest that this absorber can be effectively utilized for 5G shielding, energy harvesting, and stealth applications, offering a cost-effective and scalable solution for next-generation wireless networks.

580

Polarization-Insensitive Reflective Metalens at the Telecommunication Wavelength: Design and Numerical Study

Jagriti Ahuja

IIT Delhi

E-mail: phz208343@physics.iitd.ac.in

Here, we report the design of a reflective Metalens optimized for the mid-IR region. The Metalens is made up of coaxial Silicon nanopillars. The Si pillars stand on Aluminium film to achieve 99% reflection. The working wavelength of the Metalens is set as 1550 nm (Mid-IR region). Our simulation-driven approach successfully produced reflective Metalens optimized for the telecommunication wavelength of 1550 nm with a focal length of 200 um, a far-field efficiency of 94%, and a numerical aperture of 0.25. The proposed reflective Metalens design holds great promise for applications in mid-IR imaging, chemical sensing, beam shaping, and optical communication. Its compact and efficient nature addresses the limitations of conventional lenses in this wavelength range.

588

A 1-Bit Reconfigurable Reflective Metasurface for Sub-6 GHz Applications

Vaibhav Srivastava[1*], Debalina Ghosh[1] & Sankarsan Padhy[2]

[1]*IIT Bhubaneswar*

[2]*PXE Chandipur*

E-mail: s23ee09001@iitbbs.ac.in; deghosh@iitbbs.ac.in; s.padhy.pxe@gov.in

This paper presents a simple reflective metasurface design which can dynamically control the direction of reflected beam by changing phase states across the aperture of the metasurface. Single PIN diode is used in each unit cell (1-bit coding) to change the phase states between 0° and 180°. An 8×8 array created with the designed unit cell is updated with a suitable coding pattern calculated using Array Factor method. Simulation result shows that the designed reconfigurable reflective metasurface can steer the reflected beams up to $\pm 45°$. The proposed design can be used for sub 6GHz applications cantered at 5.8 GHz.

Contributory Sessions 15

Session SU.15.H2.X

DATA PROCESSING AND AI FOR IMAGING

274

A Zero-Reference based HE + Deep Enhancement Approach for Low-Light Images

Abhipsa Mahala[1*], Ashish Ranjan[1], Prabhat Dansena[1] & Gautam Kumar[2]

[1]*C. V. Raman Global University, Bhubaneswar*
[2]*National Institute of Technology Delhi*
E-mail: abhipsa.mahala@cgu-odisha.ac.in; ashish.ranjan@cgu-odisha.ac.in; p.dansena23@gmail.com; gautam@nitdelhi.ac.in

Low-light images are ubiquitous during night surveillance, military operations, automobiles night vision, etc, limiting the ability to perform image-based tasks. Deep learning for low-light images, although successful, still faces tough challenge dealing with uneven lighting conditions. In this work, a hybrid solution is introduced to enhance the lighting conditions of low-light images. The proposed solution uses image histogram and CNN based "Improved ZeroDCE++" architecture to generate enhanced quality images. Notably, "Improved ZeroDCE++" is light-weight architecture (with ≈ 11K parameters) that is based on the estimation of light enhancement curves. The experimental results using multiple datasets when compared to ZeroDCE [3] and ZerDCE++ [4] produced better results. Importantly, the proposed approach is able to handle the uneven lighting condition in an image in a much better way.

301

Metaheuristics-based Parameter Optimization of Bilateral Gamma Correction for Image Enhancement Applications

Simi R[1*], Justin Joseph[2], Vipin Venugopal[3] & Rashmi R[1]

[1]*Manipal Institute of Technology Bengaluru*
[2]*Centre for Brain Research, Indian Institute of Science*
[3]*Amrita Vishwa Vidyapeetham, Coimbatore*
E-mail: simi.vr@manipal.edu; josephjusti@gmail.com; vipinscms@gmail.com; r.rashmi@manipal.edu

Bilateral gamma correction (BGC) is a grey-level transformation for image enhancement applications. The shape of the BGC transformation curve and the perceived quality of its output depend heavily on the selection of the gamma parameter (GP) and linear combination weight (LCW). We propose a metaheuristics-based framework to automate the selection of the GP and LCW of BGC. The framework uses the VIF image quality metric as fitness and the particle swarm optimization algorithm (PSOA) for tracking the global minimum point of fitness. We observed that the BGC driven by optimal parameters derived from the VIF fitness produces outputs in which the tissue contrast is augmented well, and inherent brightness level is unruined and the PSOA converges to the highest VIF value compared to coati, salp swarm, whale, and grey wolf optimization techniques.

311

DeepDementia: A Deep Neural Network Integrated Convolutional Block Attention Module for Multi-Stage Dementia Classification from Brain Imaging

Abhijay[1], Rakesh Chandra Joshi[1*], Manan Vangani[1], Abhishek Sengupta[1], Vinay Kumar Pathak[2] & Malay Kishore Dutta[1]

[1]*Amity University, Noida*
[2]*Chhatrapati Shahu Ji Maharaj University Kanpur*
E-mail: abhiashmani@gmail.com; rakeshchandraindia@gmail.com; manan.vangani@s.amity.edu; asengupta@amity.edu; vinay@vpathak.in; malaykishoredutta@gmail.com

Dementia is a progressive neurological disorder affecting, decision-making capabilities, memory and cognitive functions. Early detection and precise staging of dementia are crucial for timely medical intervention and management. This study proposes DeepDementia, a deep learning-based classification model utilizing a Convolutional Block Attention Module (CBAM)-integrated deep neural network to identify five stages of dementia: non-demented, very mild demented, mild demented, moderate demented, and severe demented. The dataset contained MRI brain scans and standard preprocessing steps were applied prior to training of model. The CBAM module improves the extraction of features by selectively focusing on spatial and channel-wise attention. Through this mechanism, the network can focus on diagnostically significant areas while suppressing non-focal features, thus leading to better classification performance. The proposed model achieves 95.71% accuracy on the test set and 96.79% validation accuracy, outperforming previous studies that only considered four dementia stages. The inclusion of CBAM in CNN architectures shows the validity of its application to the refinement of decision-making areas, thus improving its performance and resource consumption. This study contributes to advancing automated dementia detection by introducing an attention-based framework for fine-grained classification, enabling precise staging and aiding in clinical decision-making.

373

Advancements in Image Captioning: A Comparative Analysis of Embeddings, Attention Mechanisms, and Transformer-Based Models

Venkatesh A*, **Prithvi Raj M**, **Janarthanan T** & **Sivasankar Elango**

National Institute of Technology, Trichy

E-mail: venkatesh77official@gmail.com; prithviscbpr@gmail.com; janarthanantj0@gmail.com; sivasankar@nitt.edu

The rapid proliferation of digital images necessitates efficient methods for automatic image captioning, a multimodal task that integrates computer vision and natural language processing. This study explores three key advancements in image captioning: (1) the impact of different word embeddings—GloVe, FastText, and BERT—on caption quality, (2) the role of attention mechanisms in improving the alignment between image regions and textual descriptions, and (3) the superiority of transformer based models over conventional methods. Our experimental results reveal that attention mechanisms significantly improve the relevance of generated captions, and the transformer-based architectures outperform RNN-based models by efficiently capturing long-range dependencies and generating more accurate, context aware descriptions. These findings underscore the potential of transformers in advancing image captioning systems, paving the way for improved applications in accessibility, human-computer interaction, and automated content generation.

399

Enhancing Intracranial Haemorrhage Classification using Feature Fusion in Brain Stroke Imaging

Thireesha Suryadevara, Mudassir Rafi* & Naveenkumar Mahamkali

SRMAP

E-mail: thireesha_suryadevara@srmap.edu.in; mudassir.r@srmap.edu.in; naveenkumar.m@srmap.edu.in*

Intracranial haemorrhage (ICH) classification is a critical task in brain stroke diagnosis, as timely and accurate identification can significantly impact patient outcomes. Computed tomography (CT) scans are widely used for detecting haemorrhages, but traditional classification approaches often rely on single-feature extraction techniques, which may not capture the full complexity of the medical images. To address these limitations, this study explores a feature fusion-based approach that combines multiple texture and frequency-domain descriptors. Specifically, Local Binary Pattern (LBP), Histogram of Oriented Gradients (HOG), Wavelet Transform, Scale-Invariant Feature Transform (SIFT), and Weber Local Descriptor (WLD) are employed to extract diverse and complementary information from CT images. These extracted features are then fused and utilized by advanced machine learning classifiers to improve classification performance. Experimental results on benchmark CT brain image datasets demonstrate that the wavelet-based features, when integrated with other fusion counterparts, yield the highest classification accuracy. The proposed feature fusion framework enhances the discriminative power of the model, leading to improved robustness and reliability in haemorrhage classification. This study highlights the effectiveness of combining multiple feature descriptors for medical image analysis and contributes to the development of more accurate and interpretable computer-aided diagnostic systems for stroke assessment.

436

Exploring Geospatial Data: Fundamentals and Challenges in Change Detection

MINAL APSANGI

Nirma University

E-mail: 23ptphde270@nirmauni.ac.in

The quick evolution of geospatial and satellite technology has greatly revolutionized numerous disciplines, affecting human life and decision-making. These technologies have made it possible to acquire and examine extensive geographic data, which has paved the way for climatic change, healthcare, disaster management, and others. This paper examines the fundamental components of geospatial data, such as its types, and acquiring methods, and focuses mainly on Deep learning Based change detection and challenges related to them. The goal is to point out the utility of geospatial information to resolve challenges faced by societies in today's era and highlight challenges and future directions.

514

Image Compression using Neural Networks

ABHINAV SHARMA* & SAMARJIT DAS

IIT Guwahati

E-mail: abhinavs.2016@iitg.ac.in; dass@iitg.ac.in

The importance of Neural network lies in its applications in diverse areas such as data compression, healthcare, etc. In this paper, we explore neural networks for data compression, focusing on Convolutional Autoencoders (CAEs). Using the MNIST dataset, a CAE reduces the dimensions of the image to a compact latent space, where the encoder extracts critical features for accurate reconstruction. This method improves storage efficiency while preserving essential image details. The model achieves high compression ratios with minimal loss of quality, outperforming JPEG. By comparing original, JPEGcompressed, and reconstructed images, the trade-offs between compression and visual fidelity are demonstrated. The ability to retain key features while reducing redundancy makes this approach highly effective for real-world applications. A key advantage of this method is its adaptability compression levels and image quality can be fine-tuned by adjusting hyperparameters like latent dimensions. The results highlight the potential of neural networks for efficient image compression and suggest broader applications in handling other data types, such as text, audio, and video. We believe that this method can be used for compressing images to a very large extent when compared to the standard JPEG compression while retaining the quality of the image.

549

Offline Signature Verification using CNN and Siamese Networks: A Comparative Analysis

Snehasis Nayak*, Amarnath Mohanty, Sambit Kumar Nayak, Jagannath Mondal, Harshvardhan Sharma & Raghunath Dey

KIIT University

E-mail: 21051768@kiit.ac.in; 21051626@kiit.ac.in; 21052786@kiit.ac.in; 2105546@kiit.ac.in; 21051820@kiit.ac.in; raghunath.deyfcs@kiit.ac.in

A major task in biometric security and authentication is Offline Signature Verification (OSV), which also consumes huge human effort; hence, how can we automate the validation of handwritten signatures? In order to enhance the signature verification challenge, this paper used a range of techniques and signature datasets, including a Siamese neural network and a histogram of gradient-based networks. Less overfitting is indicated by a CNN dataset splitting strategy with 92.59% accuracy. Moreover, CNN-HOG, which makes use of HOG as an input, achieved a 92.42% accuracy. It shows why feature extraction is better than usual neural network techniques. Having 98.53% accuracy, the Siamese neural network model outperformed on the 4NSignComp 2010 dataset. This work concentrated on creating highly strong techniques for problems involving signature verification and authenticity.

Contributory Sessions 16

Session SU.16.H3.X

ADVANCES IN OPTICAL COMMUNICATION

488

Implementation of SDON for Industrial Internet and Industry 4.0 in Google Colab

Sadia Sara*, Roshni Das, Sambit Bera, Swarnava Bhattacharya & Chandi Pani

Meghnad Saha Institute of Technology

E-mail: sadiasara34@gmail.com; das.roshni1707@gmail.com; sambitbera2003@gmail.com; swarnavabhattacharya443@gmail.com; cpani@msit.edu.in

The rising demand from Industry 4.0 and the Industrial Internet requires efficient, high-speed, and large-scale data transfer solutions. Traditional optical networks, with their static architectures, are often unable to dynamically adapt to changing traffic conditions. To address this challenge, this paper proposes the implementation of a Software-Defined Optical Network (SDON) to optimize network management, improve resource allocation, and enhance performance in large-scale and high-speed data transmission. This paper aims to implement an SDON-based network model in Python to support high-capacity, real-time data transmission, making it ideal for data center networks and industrial applications. Simulated results ensure better performance in terms of delay, throughput, and packet loss for an IP-based framework. This result also certifies the framework's scalability which best suits the Industrial Internet application.

538

Study on Photonics-Based Microwave Signal Generation Scheme for 5G Application

Shreepreet Sahu* & Prasant Kumar Sahu

Indian Institute of Technology Bhubaneswar

E-mail: a24ec09008@iitbbs.ac.in; pks@iitbbs.ac.in

In the context of high-speed and low-latency-based communication requirements, photonic generation of microwave signals is a key enabling technique for 5G applications. This paper explores microwave signal generation using photonics techniques by considering two modulation schemes, i.e., (i) amplitude modulation (AM) and (ii) double sideband suppressed carrier (DSBSC), using a delay interferometer (DI) and/or fiber Bragg grating (FBG) to generate high-purity microwave signals. Using two continuous wave (CW) lasers and a $LiNbO_3$ Mach-Zehnder Modulator (MZM), the proposed system produces 5G bands at 3.8 GHz, 28 GHz, 36 GHz, and 60 GHz. Spectral purity, signal clarity, and sidelobe power level variations are used to assess performance. Low phase noise and good spectral efficiency are two of the main benefits of generating microwave signals with a photonics scheme. The suggested plan is examined theoretically and verified via OptiSystem 22 simulations. The results demonstrate the system's capacity to adjust to the intended 5G frequency band generation and the quality of the microwave signal produced by the photonics approach.

550

Performance Studies of Inter-Satellite Optical Wireless Communication Links Using PPM

Nalluri Satwika* & Prasant Kumar Sahu

Indian Institute of Technology Bhubaneswar

E-mail: 21ec01007@iitbbs.ac.in; pks@iitbbs.ac.in

Inter-Satellite Optical Wireless Communication (IsOWC) is a promising technology for high-speed data transmission between satellites. This research focuses on establishing an IsOWC link between Low Earth Orbit (LEO) and Geostationary Earth Orbit (GEO) satellites using Pulse Position Modulation (PPM) sequence coding in Optisystem software. The optical wireless channel serves as the transmission medium, ensuring efficient and secure data exchange. To analyse signal enhancement, two key techniques: Single Input Single Output (SISO) and 4×4 Wavelength Division Multiplexing (WDM) scheme are employed. The PPM sequence is generated using Return-to-Zero (RZ) and Non-Return-to-Zero (NRZ) pulse generators, allowing a comparative evaluation of the Q-factor for both SISO and 4x4 WDM systems. Additionally, critical system parameters, including pointing error, transmitter aperture diameter, and receiver aperture diameter, are optimized to achieve an optimal Q-factor. This study aims to enhance the reliability and efficiency of inter-satellite communication by adopting systematically investigating the parameters using the sweep iteration in Optisystem. The findings contribute to the development of advanced IsOWC systems, paving the way for improved data transmission capabilities in space communication networks.

551

Machine Learning-Based Classification of OAM Modes in FSO Systems under Fog Conditions Using SVM and KNN

Suman Dass, Bibhudendra Acharya & Bijayananda Patnaik*

NIT Raipur

E-mail: sdass.phd2023.etc@nitrr.ac.in; bacharya.ece@nitrr.ac.in; bpatnaik.etc@nitrr.ac.in

The rapid advancement of Free Space Optical (FSO) communication presents a compelling alternative to traditional radio frequency methods. FSO technology has garnered significant momentum in recent years, driven by its capacity for high-speed data transmission, enhanced capacity, heightened security, streamlined installation processes, immunity to Electromagnetic Interference (EMI), and utilization of unlicensed spectrum. Despite these advantages, atmospheric turbulence poses a formidable challenge, manifesting in forms such as rain, fog, and haze, which can significantly degrade system performance. One of the primary issues stemming from atmospheric turbulence is fading, which detrimentally impacts FSO communication. Orbital Angular Momentum (OAM) multiplexing is a promising technique to enhance the capacity and performance of FSO communication systems. By leveraging multiple OAM modes, the system can achieve high data rates and improved spectral efficiency, making it suitable for next generation optical networks. However, accurately distinguishing between different Laguerre-Gaussian (LG) modes in an OAM-FSO system is challenging, especially under adverse weather conditions. To address this, machine learning (ML) models, specifically Support Vector Machine (SVM) and K-Nearest Neighbours (KNN), are employed to classify received signals into distinct LG modes. The performance of both models is analysed, demonstrating that SVM significantly outperforms KNN in terms of classification accuracy and robustness. The findings highlight the potential of ML-based classification in enhancing the reliability of OAM-FSO systems, paving the way for intelligent and adaptive optical communication networks.

556

Design and Study of an OAM-based Advanced Multiplexing Technique for FSO Communication System under Turbulent Channel Conditions

Subhrajyoti Sunani[1*], Suman Malik[2], Prasant Kumar Sahu[1] & Debalina Ghosh[1]

[1]*IIT Bhubaneswar*
[2]*IIT Kanpur*
E-mail: a23ec09008@iitbbs.ac.in; sm38@iitbbs.ac.in; pks@iitbbs.ac.in; deghosh@iitbbs.ac.in

The large unlicensed spectrum and high data transmission rate of Free Space Optical (FSO) communication have drawn a lot of interest. Nevertheless, meteorological circumstances and atmospheric turbulence limit the FSO system's performance. The capacity of FSO system can be greatly increased by wavelength division multiplexing (WDM), although WDM is inadequate in extreme turbulent environment. To overcome these limitations, integrating WDM with advanced multiplexing techniques such as Polarization Division Multiplexing (PDM) and Orbital Angular Momentum (OAM) enables a more spectrally efficient system capacity enhancement, even under severe turbulence and adverse weather conditions. This paper proposes a 16-channel WDM system utilizing OAM-PDM and evaluates its performance under various turbulence conditions.

558

Study on Pointing Error and Atmospheric Turbulence in FSO for 5G and Beyond Communication

Prasant Sahu* & Debalina Ghosh

IIT Bhubaneswar
E-mail: pks@iitbbs.ac.in; deghosh@iittbbs.ac.in

Free-Space Optical (FSO) communication offers a number of advantages for which it is a promising solution for high-speed, high-capacity data transmission in 5G and Beyond networks. However, its performance is severely affected by pointing error and atmospheric turbulence, which leads to signal degradation, high bit error rate (BER), intensity fluctuations, beam wander, and attenuation.

This paper presents simulated studies on performance analysis of an FSO for 5G and beyond communication. This study aims to highlight the pros and cons of applications of FSO for 5G and beyond communication. We have simulated the transmission of quadrature amplitude modulation-Orthogonal Frequency Division Multiplexing (QAM-OFDM) signals under different atmospheric conditions, weather conditions, and different pointing error scenarios. The performance of the FSO system is investigated for different weather conditions. The system performance was analysed based on bit error rate (BER), quality of eye diagrams, and standard Q-factor. The FSO distance in clear weather and in heavy fog condition found to be approximately 4km and 1 km, respectively. Whereas the system performance degrades significantly under cirrus cloud conditions where the range has fallen up to 50 meters. The system performance is analysed for different misalignment, different weather conditions and for atmospheric turbulence conditions. The proposed system is efficient enough for high data rate transmission along with high low BER for different environmental scenarios.

559

Free-Space Optical Communication: Navigating the Potential of Photon Entanglement and 16-QAM

Prasant Sahu*, Debalina Ghosh & Subhrajyoti Sunani

IIT Bhubaneswar

E-mail: pks@iitbbs.ac.in; deghosh@iittbbs.ac.in; a23ec09008@iitbbs.ac.in

Free-Space Optical (FSO) communication technology can cater to various types of connectivity requirements within optical networks, including core, edge, and access networks. However, the performance of FSO communication links can be adversely affected by severe atmospheric conditions, such as turbulence and weather-related factors. In this study, we have investigated the benefits of photon entangled 16-QAM- FSO link in improving the performance of FSO communication under diverse weather and atmospheric turbulence conditions. The efficient implementation of a practical secure quantum communication network may revolutionize the terrestrial and non-terrestrial communication both form speed and security point of view.

565

Reliability of an FPGA-Based Air-to-Water-to-Air VLC Link: Experimental Investigations

Vedant Borde[1*], Saroj Mahapatra[2] & Shailendra Varshney[1]

[1]*IIT Kharagpur*

[2]*Seemanta Engg. College*

E-mail: vedant.borde.24@kgpian.iitkgp.ac.in; skmahapatra.iitkgp@gmail.com; skvarshney@ece.iitkgp.ac.in

In this work, we demonstrate the reliability of a visible light communication (VLC) link through the air-to-water-to-air interfaces considering various channel-based impairments, both in dark and bright environments. The experiment result shows that, along with the generated salinity and water bubbles in dark and bright environments, the refractive index variation has a significant impact on the VLC system's performance. The current work contributes to optimizing VLC-based systems for enhanced reliability in dynamic underwater and air environments, aiding in the development of robust optical communication links for real-world applications.

Contributory Sessions 17

Session SU.17.H4.X

SENSING FOR HUMAN MACHINE INTERFACE

128

Kannada Speech to Sign Language Converter: A Novel Platform Integrates Indian Sign Language and Speech-to-Text Technology

Pushpa Latha*, Parkavi A, Mushtaq Ahmed D M, Arkodeep Koley, Ankesh M Naik & Ashish Kumar

Ramaiah Institute of Technology

E-mail: pushpalathamn@msrit.edu; parkavi.a@msrit.edu; mushtaq.ahmed@msrit.edu; 1ms20is022@msrit.edu; 1ms20is015@msrit.edu; 1ms20is027@msrit.edu

India faced an enormous challenge in providing accessible educational resources for its deaf community, particularly in rural areas where specialized schools and sign language interpreters were scarce. An estimated 18 million people suffered from hearing impairments, many of whom were children with severe to profound hearing loss, highlighting the critical need for innovative solutions. Researchers addressed this pressing issue by introducing a novel online platform designed to close the accessibility gap for the deaf community in terms of communication and education. The software utilized technology to link visual sign language cues with Kannada speech-to-text conversion, providing autonomous sign language learning and communication. This approach held great promise for augmenting accessibility and empowering members of the deaf community. The program translated spoken Kannada into sign language motions using Google API services, such as Translation, Speech-to-Text, and Avatar. This creative solution promoted social inclusion, empowered deaf people, and increased awareness of the needs and skills of the deaf community by facilitating effective communication between those who were not familiar with sign language and the deaf population.

149

Model Agnostic Few Shot Learning for Finger Vein Recognition

Kiranjit Pattnaik*, **Partha Pratim Sarangi** & **Bhabani Shankar Prasad Mishra**

KIIT University

E-mail: kiranjit.pattnaik99@hotmail.com; ppsarangi@gmail.com; mishra.bsp@gmail.com

In the field of biometrics and medical imaging, research on finger vein biometrics has been active and effective with limited datasets that are publicly available to perform experimentation. However, with how secure and reliable the finger vein images are used and with limited set of examples or users with limited samples there is a high need to create solutions to tackle the classification of limited users which would be used for limited accesses. In this paper we have focused on building various ensemble techniques along with few shots learning methodology which is model agnostic and would give the best accuracy in any possible scenario by using convolutional neural network (CNN) and Siamese network along with usage of the weights of couple of pretrained image models. Three different approaches with fusion mechanism have been proposed to predict better even with very few examples of the data. The various performance metrics, hyperparameter tuning, feature extraction, tables and figures are explained in the paper.

182

Object Detection for BVI People in Low Light Conditions using Improved TBEFN Technique

Shraddha Gugulothu

YCCE

E-mail: shraddha11sangewar@gmail.com

Traveling and interacting with others in their environment can be extremely challenging for those who are blind or visually impaired (BVI), especially under low-light conditions when the detection of objects is further compounded. In this paper, an enhanced Two-Branch Enhanced Feature Network (TBEFN) specifically formulated for low-light image enhancement in order to make object detection easy for BVI users is proposed. The system incorporates leading-edge image processing with a stress on boosting visibility in adverse light conditions. Enhanced TBEFN utilizes two-branch processing in order to double up and jointly boost global contrast and local information, such that the output remains better balanced as well as more perceptually sensitive. Low-light datasets undergo a thorough testing on which baseline models are seen to improve greatly regarding detection accuracy as well as image clarity.

217

Efficacy Prediction of Low Observable Measures with Range & Search Time Using Eye Tracker

Amit Bhati*, Rahul Srivastava, Sohan Lal Borana & Vijay Singh Choudhary

Govt

E-mail: amitbhati.dl@gov.in; rahulsrivastava.dl@gov.in; sl.borana.dl@gov.in; vs.choudhary.dl@gov.in

Low Observable Measure (LOM) plays a decisive role in the signature reduction of the target to increase its survivability. The LOM attempts to reduce the contrast of the target so that it blends in the immediate background. Consequently, LOM creates a state of confusion during the target search-cum-acquisition process which results in enhancing the survivability of the target. The performance of LOM is measured in terms of reduction in target detectability. Two methodologies namely subjective & objective methods are reported in the published literature to quantify the performance of the LOM. The subjective approach is based on observer response whereas the objective approach applies image processing, artificial intelligence, and other techniques to quantify the effectiveness of the LOM. Both the subjective & objective methods quantify the performance of the LOM in terms of reduction in target detectability, particularly with range. In the target search process, the search time is a significant parameter. Therefore, it is imperative to estimate the effectiveness of the LOM in terms of search time during the target detection process. The work presented in the article aims to gauge the performance of the LOM in terms of search time. An eye tracker device was employed to measure the search time parameter during the target acquisition. The process steps involved in the data collection and conduct of photo simulation activity using eye tracker are discussed in detail. The photo-simulation activity using an eye tracker was conducted with thirty-five observers to deduce the target detectability with time. The eye tracker also facilitates to log the observer response in the form of mouse click. The mouse click enables the calculation of the detection probability. Thus, it is viable to characterize the performance of LOM using an eye tracker in terms of detection probability with range and search time, with minimum human involvement.

307

Contrastive Loss Coupled with Occlusion Aware Training Aids in Face Recognition from Low Quality Group Photos

Sam Narayana Subudhi[1*], Mrinal Kanti Das[2] & Muhammed Favaz[3]

[1]*Indian Institute of Technology, Bhubaneswar*
[2]*Indian Institute of Technology, Palakkad*
[3]*Samsung Research India*
E-mail: samnarayana98@gmail.com; mrinal@iitpkd.ac.in; favazkv9161@gmail.com

In scenarios without restrictions, recognizing faces from a group photo is still a difficult task. There are two major challenges: (i) the faces can be partially occluded, (ii) due to low resolution facial features may appear similar among multiple faces. State of the art methods are found to suffer in the presence of these challenges. In this paper we propose a method to mitigate this limitation. We artificially occlude training images to make the face recognition models be aware of occlusion. Secondly, we put a contrastive loss in a cohort to avoid the face recognition model being confused between similar looking faces due to low quality. In multiple datasets we have shown that state-of-the-art face recognition models improve their performance with the proposed training mechanism.

485

Person Tracking System

Keerthana P*, Chalukya Nayaka B K, Harini J, Bharath K B & Mahalakshmi J

Reva University
E-mail: chalukyanayakabk2@gmail.com; chalukyanayakabk@gmail.com; harinijanardhan03@gmail.com; bharathkb2145@gmail.com; mahalakahmij5@gmail.com

The rise in demand for secure, automated, and contactless identity verification systems has driven the need for intelligent biometric solutions. This paper presents Watchwise, a real-time face recognition system designed for person tracking and identity logging using advanced deep learning and computer vision techniques. The system employs MTCNN (Multi-task Cascaded Convolutional Networks) for accurate face detection and alignment, followed by FaceNet for extracting robust 128-dimensional facial embeddings. Identity matching is performed using cosine similarity against stored embeddings to verify individuals efficiently. To prevent unauthorized access and spoofing attempts, the system integrates an anti-spoofing module comprising blink detection, texture analysis using CNNs, and optional infrared sensing. A lightweight SQLite database is used for secure local storage of embeddings and activity logs. The system is benchmarked using a combination of standard datasets such as LFW and CelebA, along with custom datasets to evaluate performance in practical conditions. Metrics including accuracy (95.2%), precision, recall, F1-score, FAR, and FRR confirm the system's effectiveness. Furthermore, the solution is optimized for real-time operation on edge devices such as Raspberry Pi, making it suitable for resource-constrained environments. The system presents a scalable, privacy-aware solution for secure person tracking applications in educational, corporate, and security domains.

Contributory Sessions 18

Session SU.18.H5.X

BIOMEDICAL IMAGING AND SENSORS

218

CMOS Capacitor-Coupled Based Instrumentation Amplifier for ECG Signal Acquisition

Oprabhat Barik

IIT BHUBANESWAR

E-mail: pkb15@iitbbs.ac.in

A capacitor-coupled based instrumentation amplifier (CCIA) is a crucial circuit for an analog front-end ECG signal acquisition. The CCIA is an essential component for demanding in wearable and IoMT devices. On response to these needs, a low power, low input-referred noise (IRN), high common-mode-rejection-ratio (CMRR) and total harmonic distortion (THD) using UMC 0.18-μm standard CMOS technology has been designed at a transistor level. The Miller op-amp of the CCIA is designed in weak inversion region. This instrumentation amplifier is well simulated at PVT (process, supply, temperature) for its performance metrics such as closed loop gain, CMRR, IRN, THD and power consumption. The maximum power consumption is 1.736μWatt. The CCIA circuit is tested by using the Physionet ECG database. This circuit enhances the signal processing capabilities and enables accurate data acquisition.

310

Fabry-Perot based Flexible System for Human Pulse Monitoring

Pratik Mishra* & Rajan Jha

IIT Bhubaneswar

E-mail: pm43@iitbbs.ac.in; rjha@iitbbs.ac.in

Sensitive wearable technologies that can monitor vital bodily functions such as heart rate have become more relevant for both personal and public health. We demonstrate a wearable system with a flexible cavity embedded in a polydimethylsiloxane (PDMS) membrane that can detect the low frequency of 1 Hz. This ability makes the system more effective for low-frequency-based human pulse detection. Furthermore, the suggested method allows for the reliable and consistent measurement of pulse rate in a variety of locations, including the wrist, elbow, and fingertip. This system could be utilized as a wearable device for a variety of medical conditions.

364

Enhanced Multi-Class Brain Tumour Classification Model using MRI Images: A Wavelet and Deep Learned Features Based Approach

Nihar Ranjan Panda, Debendra Muduli*, Sourav Parija & Santosh Kumar Sharma

CV Raman Global University

E-mail: niharpanda1982@gmail.com; muduli.debendra@gmail.com; parijasourav3025@gmail.com; santoshku.sharma17@gmail.com

Accurate detection of brain tumours from Magnetic Resonance Imaging (MRI) scans is a critical need in modern healthcare, as early diagnosis and precise tumour typing can greatly influence treatment strategies and improve patient prognosis. In this research, we introduce a hybrid classification framework that fuses handcrafted wavelet-based features with deep CNN embeddings extracted from a pre-trained VGG19 network. After an extensive pipeline of image preprocessing, wavelet decomposition, and CNN-based feature extraction, we employ a stacked ensemble classifier composed of XGBoost, Random Forest, Support Vector Machine, and K-Nearest Neighbours, with XGBoost serving as a meta-learner to optimally combine the predictions of all base models. Our experiments on a comprehensive dataset of 1311 MRI images demonstrate a final test accuracy of approximately 92.14 %, bolstered by highly positive precision-recall metrics. Detailed discussion of the confusion matrix and classification report reveals that no-tumour and pituitary cases are most consistently distinguished, whereas glioma and meningioma classes exhibit some degree of overlap, aligning with well-known clinical challenges. Overall, these findings reinforce the viability of combining handcrafted textural features with learned deep representations for accurate and robust neuro-oncological decision support.

411

Design a Compact UWB Antenna for the Detection of Breast Tumour via Non-Invasive Microwave Imaging

Tarun Panwar[1*], Aryan Khanna[1], Manpreet Kaur[2], Asit Ranjan Mridha[3], Vishal Srivastava[1] & Hari Shankar Singh[1]

[1]*Thapar Institute of Engineering and Technology Patiala*
[2]*Mody University (MU), Laxmangarh, Sikar, Rajasthan, India*
[3]*All India Institute of Medical Sciences (AIIMS), New Delhi*
E-mail: tarunpanwar1931@gmail.com; akhanna0311@gmail.com; preetsmanu94@gmail.com; asit_aiims@yahoo.co.in; vishalsrivastava17@gmail.com; harishankar1990@gmail.com

This paper proposes a non-invasive microwave imaging method for breast cancer detection using a compact ultra-wideband (UWB) antenna. This method is safe and secure since it uses non-ionizing radiation, eliminating the possibility of dangerous radiation exposure. The antenna has overall dimensions of 12 × 12 × 1.6 mm3. The antenna operates in the frequency range of 5.38 to 15 GHz and has a peak gain of 3.59 dBi at 13 GHz. The effectiveness of the proposed UWB antenna is evaluated by creating a breast phantom that replicates a realistic scenario of a human body part using the dielectric property of the tissues in the CST MWS. This phantom has three different tissue types: skin, fat, and glandular. When the tumour tissue is solidly embedded in the glandular tissue area of the phantom, bistatic modelling is done. Furthermore, the antenna is appropriate for microwave imaging of breast cancer, according to the time-domain analysis. The tumour location

419

WatershedSE-UNet: A Hybrid Deep Learning and Morphological Approach for Lung Nodule Segmentation in CT Scans

Jiddu Krishnan O P* & Pinki Roy

NIT Silchar
E-mail: jiddu21_rs@cse.nits.ac.in; pinki@cse.nits.ac.in

Lung nodule segmentation plays a crucial role in the early detection and diagnosis of lung cancer. In this study, we propose a novel WatershedSE-UNet, leveraging watershed-based segmentation alongside deep learning-based feature extraction to enhance segmentation accuracy. Unlike conventional U-Net and DeepLab models, which primarily rely on convolutional architectures, our approach integrates morphological processing with multi-scale feature learning, enabling more precise boundary delineation of lung nodules. We evaluate our model on the LUNA16 dataset, demonstrating its effectiveness in segmenting nodules of varying sizes and intensities. The proposed model achieves a Dice Similarity Coefficient (DSC) of 83.87%, outperforming several state-of-the-art methods in terms of segmentation accuracy and precision values. Studies performed confirm that integrating watershed-based segmentation significantly improves performance by refining boundaries and mitigating false segmentations. Additionally, the model's robust learning curve suggests strong generalizability to unseen data. The results indicate that WatershedSE-UNet is a highly effective and reliable approach for lung nodule segmentation, offering potential applications in computer-aided diagnosis (CAD) systems. Future work will explore model enhancements to further improve sensitivity while maintaining precision, making it a valuable tool for radiologists and medical practitioners in lung cancer screening and prognosis.

592

Deep Spatio-Temporal Networks for Visual Speech Recognition: A Hybrid CTC-Cross Entropy Approach

Nikhil Niranjan* & Raushni Kumari

IIT Bhubaneswar

E-mail: 23sp06010@iitbbs.ac.in; 23cs06017@iitbbs.ac.in

Visual Speech Recognition (VSR) is a challenging task that involves interpreting spoken language by using visual cues from lip movements. This paper presents an end-to-end deep learning framework for lipreading, leveraging a combination of 3D Convolutional Neural Networks (CNNs) and Bidirectional Long Short-Term Memory (BiLSTM) networks. The model processes grayscale video frames, extracts spatiotemporal features using 3D CNN layers, and models temporal dependencies using BiLSTM layers. A hybrid loss function combining Connectionist Temporal Classification (CTC) and Categorical Cross-Entropy (CE) was used to improve the robustness of phoneme sequence alignment.

AUTHOR INDEX